OVERDUE FINES:
25¢ per day per item

MAB Technical Note 11

Fire and fuel management in mediterranean-climate ecosystems: research priorities and programmes

unesco

MAB Technical Notes 11

Titles in this series:

1. *The Sahel: ecological approaches to land use*

2. *Mediterranean forests and maquis: ecology, conservation and management*

3. *Human population problems in the biosphere: some research strategies and designs*

4. *Dynamic changes in terrestrial ecosystems: patterns of change, techniques for study and applications to management*

5. *Guidelines for field studies in environmental perception*

6. *Development of arid and semi-arid lands: obstacles and prospects*

7. *Map of the world distribution of arid lands*

8. *Environmental effects of arid land irrigation in developing countries*

9. *Management of natural resources in Africa: traditional strategies and modern decision-making*

10. *Trends in research and in the application of science and technology for arid zone development*

11. *Fire and fuel management in mediterranean-climate ecosystems: research priorities and programmes*

Fire and fuel management in mediterranean-climate ecosystems: research priorities and programmes

Edited by James K. Agee

unesco

Launched by Unesco in 1970, the intergovernmental Programme on Man and the Biosphere (MAB) aims to develop within the natural and social sciences a basis for the rational use and conservation of the resources of the biosphere and for the improvement of the relationship between man and the environment. To achieve these objectives, the MAB Programme has adopted an integrated ecological approach for its research and training activities, centred around fourteen major international themes and designed for the solution of concrete management problems in the different types of ecosystems.

SCIENCE LIBRARY
QH
545
.F5
F57

Published in 1979 by the United Nations
Educational, Scientific and Cultural Organization
7 Place de Fontenoy, 75700 Paris
Printed by Union Typographique
Villeneuve-Saint-Georges

ISBN 92-3-101688-1

L'aménagement des feux et des produits
combustibles dans les écosystèmes à climat
méditerranéen: priorités et programmes de
recherche: ISBN 92-3-201688-5

© Unesco 1979
Printed in France

Preface

Fire exerts a strong effect on plant communities and landscapes in a number of regions of the world - none more so than those regions having a mediterranean climate. Fire has long been the simplest, quickest and most economic way of destroying undesirable natural vegetation and clearing space for pastures and crops. Since Paleolithic times, it seems that the shepherds and farmers of the countries around the Mediterranean Sea have periodically burned the forests to create pastures or new agricultural land. Intentional burning remains an important management tool in many mediterranean-climate countries.

At the same time, wildfires sweep over several millions of hectares of land every year. The social, economic and environmental effects of such fires can be devastating. In 1970, for example, wildfires swept over 160,000 ha in Spain and southern France alone. In the same year, in California, a single wildfire burned for thirteen days, scorched more than 200,000 ha of land, killed sixteen people, destroyed 722 homes, and cost more than a quarter of a billion dollars in direct suppression costs and structural losses.

Approaches to controlling wildfires have varied. Some areas have no control policies. Others have management policies ranging from total fire exclusion to controlled burning to fuel manipulation. The environmental costs of these management possibilities have never been fully assessed. In those areas where a clear policy is evident, it may not be carried out because of social or fiscal constraints, or because of lack of information transfer between researchers and resource managers.

The need to develop an understanding of the full dynamics of fire-type ecosystems of mediterranean regions is urgent so that management policies can be developed and assessed on a rational basis. Ways must be developed to facilitate communication between researchers and resource managers facing similar problems in different parts of the world. Mechanisms must be developed to translate research findings into practical programmes that can be transmitted quickly to resource managers and the public.

Recognition of these needs and objectives led to the convening in 1977 of an international MAB symposium on the 'Environmental consequences of fire and fuel management in mediterranean-climate ecosystems'. The symposium was held at Stanford University, Palo Alto, California, from 1 to 5 August 1977. The symposium was jointly organized and sponsored by the United States Government as well as the Scientific Committee on Problems of the Environment (SCOPE), the Jasper Ridge Biological Reserve (Stanford University), with the support of Unesco-MAB and the US National Science Foundation.

The symposium was attended by about 200 participants, from some 8 countries. It was divided into six sessions covering: the nature of the world's mediterranean ecosystems; an assessment of man's interactions with those systems; regional problems and approaches; contributions to the study of mediterranean ecosystems; identifying research problems; and a field trip to observe management practices in forests and scrublands of California.

The proceedings of the symposium have been edited by H.A. Mooney and C.E. Conrad and published in 1977 by the Forest Service of the United States Department of Agriculture, as General Technical Report WO-3. The proceedings volume contains sixty presentations, covering four main topics about the environmental consequences of fire and fuel management, namely: role of fire in regions with a mediterranean climate; interaction of man and fire in these regions; problems in fire and fuel management and regional approaches to their solution; and contributions toward the study of mediterranean ecosystems, including chaparral and conifer systems.

Workshops convened during the symposium identified priorities for future research, including the development of international cooperation. This MAB Technical Note, divided into six sections, attempts to distill the principal technical findings presented at the symposium, and the main recommendations agreed upon by participants.

A first section deals with species characteristics of plants in fire-prone mediterranean climates, and the importance to management of knowledge on the response of plants to different fire regimes. Priorities for research include identification of the past regimes of 'natural fires', assessment of factors affecting flammability, response of invader plant species to fire, and the productivity of shrublands under varying intensities and frequencies of burning, grazing and browsing.

A second section is concerned with the ecological impact of fires on pools and fluxes of minerals and sediments in mediterranean ecosystems. Five research priorities are considered to be of particular importance, namely the sizes of nitrogen and phosphorous pools, the patterns of cycling of nitrogen and phosphorous, the development of systems models of nutrient cycling, plant adaptations to fire as the primary agent of decomposition, and the significance of fire in promoting erosion and other geomorphological changes in pools of sediments.

The ecological impact of fires on the productivity, structure and diversity of plant and animal communities forms the focus of the third section of this Technical Note, and specific recommendations are made for research in these subject areas. It is also suggested that long term demonstration and research plots be established under different conditions and degrees of human impact in all of the mediterranean climate regions of the world.

A fourth section deals with methods for assessing alternative policies for the management of fire. The needs for analytical models which can provide conceptual frameworks for research are discussed. Research directions are identified for helping improve the various strategies that comprise most fire policies in current use; these strategies include fire suppression, prescribed fire and natural fire management.

A fifth section is focussed on management problems and solutions at the interface between man and mediterranean wildlands. Four problem areas are identified as high priority for future action: pilot projects for integrated research and for demonstrating the application of all existing and relevant knowledge on fire management; research on fire prevention; dissemination of information, and training on the fundamentals of the behaviour of prescribed fires; procedures for innovative fuel management in different mediterranean regions.

The problems of improving information flow between research workers, resource managers, and the general public are dealt with in a final section of this Technical Note. Participants at the Stanford symposium proposed a series of measures to enhance communication between and among the various groups concerned.

The concern of the symposium participants toward solving the wildfire problem promoted them to adopt unanimously the following resolution:

Throughout mediterranean-climate countries of the world, the peoples of different nations have problems in managing wildland vegetation to conserve and enhance their resources as well as to contain wildfires. Traditional programmes focussing on fire suppression do not consider the relationship of fire to the environment.
Research and practice have shown that wildland-fuels management can maintain and improve resources. In accordance with changing land-use patterns, such management should recognize all tools available, including prescribed fire. The public will benefit from wildland management on both public and private lands and should, therefore, share the costs and risks.
This symposium recommends that governments designate pilot areas to implement and demonstrate wildland-vegetation management programmes as well as for the study and review of the role of fire in ecosystem and resource management.

In publishing this Technical Note, Unesco hopes that the information and recommendations presented therein will be of use and interest to scientists and resource managers in all countries where fire has a major impact on the environment, and on the livelihood and well-being of the local populations concerned.

It is recognized that much of the material presented in the Note is based on mediterranean climate ecosystems in California, and that somewhat less attention is given to other mediterranean climate zones. For these zones, and particularly for the countries of the Mediterranean Basin, attention is drawn to a number of complementary publications and documents that have been produced within the framework of MAB. These include MAB Technical Notes 2 on 'Mediterranean forests and maquis: ecology, conservation and management'.

Reference is also made to the reports of several regional MAB meetings which have included discussion on the effects of fire in mediterranean areas. These meetings have included: the regional meeting for the arid and semi-arid zones of North Africa, held in Sfax, Tunisia, in April 1975 (see MAB Report Series No. 30); the regional meeting for the northern Mediterranean countries, held in Potenza, Italy, in October 1975 (see MAB Report Series No. 36); and the MAB Mediterranean Scientific Conference, held in Montpellier in September-October 1976 (see MAB Report Series No. 43); mention should also be made of the FAO/Unesco Technical Consultation on Forest Fires in the Mediterranean Region, held in St. Maximin in May 1977.

Unesco wishes to thank all those who collaborated in convening the Stanford symposium, including the various sponsoring organizations and the Steering Committee of the symposium, which comprised Harold A. Mooney, C. Eugene Conrad, Bernard Kilgore, and John W. Menke. Unesco also wishes to express its thanks to those who have contributed to the preparation of this MAB Technical Note. In addition to the authors of the different sections of the Note, special thanks are due to Dr. James K. Agee of the University of Washington, who was responsible for editing the text of the Note. The material, the points of view and the opinions expressed in this publication are the sole responsibility of the authors and are not necessarily shared by Unesco.

Contents

The role of species characteristics of plants in fire climates
as a guide to management, by A.M. Gill 11

Ecological impact of fires on mineral and sediment pools
and fluxes, by P.W. Rundel 17

Ecological impact of fires on community productivity, structure
and diversity, by J.W. Menke 23

Alternative fire management policies: methods for assessment,
by J.K. Agee 27

Management problems and solutions at the interface between man
and mediterranean wildlands, by C.C. Wilson. 33

Facilitating communication between researcher, manager and
the public, by A.R. Taylor and R.Z. Callaham 39

The role of species characteristics of plants in fire climates as a guide to management

A. Malcolm Gill

INTRODUCTION

Various combinations of fire events can have different effects on landscapes according to the fuel type, the interval between events, the times of year at which they occur and the intensity of each occurrence. Some of these combinations can enhance the chances of survival and reproduction of a plant species while other combinations can lead to the demise of the same species. Such species responses to different fire regimes, and the variation of responses within a plant community, can allow a manager to manipulate the composition and character of vegetation to achieve certain management objectives.

In different landscapes, the objectives of management can vary widely. The manager of a national park may wish to restore a landscape to its 'natural' state after a long period of fire suppression; a forest manager may wish to optimize timber production and minimize occurrences of high intensity fire; a manager of a floral reserve may wish to enhance populations of rare species. The alternatives are numerous and the interested reader may find Kayll (1974) a useful reference on this subject. This section of the Technical Note seeks to identify some major areas for research related to a limited number of management objectives and involving the study of the responses of individual species to various combinations of fire events.

While there are certain advantages and disadvantages in using this method to study the effects of fires, it is by no means the only method of approach. In botanical studies, at least, three broad methods of the study of fire effects can be identified, including the species-response method (Gill 1977a). The other two major approaches are what could be termed the 'stand-ordination' approach and the 'fire-regime' approach. The ordination approach in its simplest application consists in the comparison of two stands of vegetation, burned at different times in the past; the approach reaches its greatest sophistication when a gradient of time intervals since burning is considered along with gradients of other environmental influences (see, for example, Kessell 1976). The fire-regime approach is an experimental approach by which fire regimes are imposed on pre-measured vegetation. This approach has been used particularly in Africa during the last fifty years.

SPECIES-CHARACTERISTICS APPROACH AND THE MANAGEMENT OF FIRE-PRONE LANDSCAPES

The species-characteristics approach uses the attributes of a species as a guide to behaviour. This approach has recently been used, for example, in the modelling of plant succession after fire by Noble and Slatyer (1977) who identified a number of 'vital attributes' of species important to behaviour in fire-prone situations. Historically, it has been used indirectly in the formulation of management attitudes to fires, and directly in the management of, for example, landscapes for forestry, wildlife and plant species conservation.

That management attitudes to fires affect management action is well known in this era of changing management practice. This has become clear in the United States in particular, where long-standing attitudes about total fire suppression are being modified to include 'limited suppression' fire policies. Attitudes are changing because there is a new appreciation of fire effects and behaviour.

Observations of certain species characteristics, such as stimulation of fruit dehiscence by fire, enhancement of seed germination by fire, promotion of flowering by fire, and survival of plants after burning, have led some observers to conclude that management fires will not harm the vegetation resource because the flora are 'adapted to fire'. This management attitude is an unfortunate one because it is inaccurate. A truer statement would be that 'flora can be adapted to certain fire regimes', a statement which recognizes the need to consider species characteristics within the framework of the natural or imposed fire regime in order to ascertain species responses (Gill 1977a).

Direct applications of the species-characteristic approach to management have been made in the practice of forestry. For example, in eucalypt forests of Australia valuable as a timber resource - such as the forests of *Eucalyptus regnans* F. Muell. (family Myrtaceae) in the South East - silvicultural systems seem to have evolved in response to a knowledge of the species responses to fire regimes (e.g. from Galbraith 1937, Jacobs 1955, Gilbert 1959). The dominant eucalypts, in natural systems, seem to have regenerated only after infrequent fires which have had the effect of killing the dominant trees, removing a dense suppressive understory of non-eucalypt species, and allowing seed (stimulated to fall from capsules on the tree by the fire) to germinate readily in a well-lit, nutrient-enriched seedbed from which the competitive understory has been removed. In the silvicultural systems, the trees are logged, the understory and logging debris burned, and seedlings established either from planted seedlings, sown seed, or seed shed from trees left standing as a seed source. In Australia, recognition of the rôle of fires in regeneration of the dominant species seems to have led to successful silvicultural practice in that country's most productive timber areas.

For wildlife, a knowledge of the animal's diet and shelter requirements, together with an understanding of plant-species characteristics can be useful in developing suitable fire-management policies. This is particularly applicable when there is a close dependence of an animal on a species of plant markedly influenced by fires. An example is the grouse (*Lagopus lagopus scoticus*) dependent on the brush *Calluna vulgaris* (L.) Hull (family Ericaceae) in Britain. This plant, and the bird population dependent on it, is maintained by management fires at 10-15 year intervals (Gimingham 1971). This regime, while optimal for the grouse and heather, is disadvantageous for a number of other species and had led to a virtual monoculture of *Calluna*. Gimingham (1971) notes that 'unfortunately, even the normal 10-15 year burning rotation is unfavourable for wildlife generally, and therefore to some extent incompatible with nature conservation. Good burning management - [for grouse, etc.] ... is associated with a restricted and monotonous flora and fauna...'. This example emphasizes the importance of management objectives in assessing the desirability of fire-regime effects. For grouse management the regime is excellent, but for flora and fauna conservation other regimes could be more desirable.

For plant-species conservation, knowledge of life histories and species characteristics can allow the critical limits for survival and reproduction to be known and thus provide a guide to management. The giant sequoia (*Sequoiadendron giganteum* Lindl. - family Taxodiaceae) of California provides an example: 'The current management programme... is based on the hypothesis that the giant sequoia exists today because of the rôle fire has played in its life cycle' (Kilgore 1973). Other examples can be found where population enhancement is desired in order to improve floral displays for tourism, conserve dwindling populations of a rare species or to improve the ability of native species to compete with invasive exotics.

These examples illustrate some of the uses of the species-characteristic approach in formulating management plans for fire-prone landscapes. Many more examples could be given of other rôles that such an approach can have in fire management programmes. Species characteristics can be used by management to illustrate the rôles fires have had and can continue to have in natural landscapes, and therefore be of use in national park interpretive programmes.

PRIORITIES FOR RESEARCH

At the Stanford workshop, broad areas were distinguished in which some specific research projects could be identified. The research topics are not listed in order of priority. Documentation of particular examples is necessarily brief because of the information available. For each problem area, an indication is given of how the species-characteristic approach can be useful to the solution of these problems, and of how some of these studies could be approached. The workshop considered that a knowledge of life history attributes was important in understanding ecosystems and their management, although it was at the same time recognized that there are a diversity of management objectives. It was also considered that this approach could be of value even where management objectives centred on collective properties of vegetation such as fuel loading, because species compositions of otherwise similar vegetation can have important consequences for vegetation behaviour.

Emerging from the discussions in the workshop was a desire to classify species responses according to life history attributes. How this could be done was not discussed in detail, but it was thought desirable to be able to predict species responses to fire regimes. As various classifications already exist in the literature, such a task could probably be effectively accomplished through literature review, at least in the first instance.

Identification of 'natural' fires regimes

Many managers of national parks in the United States are attempting to discover what the 'natural' fire regime was prior to white occupation, in an effort to maintain ecosystems in 'as near pristine conditions as possible' (e.g. Houston 1971). Some of the methods being used are the study of fossil charcoal deposits (e.g. Byrne, Michaelsen and Soutar 1977) and fire scars on trees (see Houston 1971), and the analysis of differences between historical and modern photographs (see van Wagtendonk 1974). While valuable in some areas, these methods are not always appropriate because of the absence of suitable material. Using species characteristics, further information on 'natural' fire regimes can be gathered.

The season of year in which burning is 'natural' can be a contentious issue where prescribed burning is implemented. Information on how species respond to the time of year in which burning occurs is beginning to appear in the literature. Enhancement of populations by burning at certain times can indicate a natural season of burning, while a decline of populations caused by burning at other times of year can indicate an artificial situation.

Kruger (1977a) has data on the South African geophyte *Watsonia pyramidata* (Andr.) Stapf. (family Iridaceae) which illustrate the effects of season of burning. In unburned populations, about 5 per cent ramets flower annually. If burned in the spring, this situation remains unchanged; if burned in autumn and possibly late summer, however, about 50 per cent ramets produces inflorescences. Seasonal flowering responses have also been recorded in ongoing work reported by Gill (1977b) for a suffrutescent monocotyledon in Australia (*Xanthorrhoea australis* R. Br., family Xanthorrhoeaceae). Leaf removal in spring and summer caused abundant production of inflorescences, while autumn and winter treatments had produced relatively few inflorescences 6 months after final treatment. Taylor (1977) provides another very interesting example of the interaction of season of burn, seed production and fungal infection in determining the regeneration of the shrub species, *Staavia dodii* Bolus (family Bruniaceae). To maintain the species, fires during the period from late autumn to late spring seem most appropriate.

Fire intensities of the past are difficult to discern but some data are available from a knowledge of species characteristics, particularly those of trees. Old even-aged stands of trees may have had a common origin after fire, and intensities necessary to kill such trees may be predicted. Fire scarring could also be used to suggest fire intensities (as well as frequencies) as it is possible to interpret the fire resistance of the tree at the time of first scar formation from the size of the tree (shown by growth rings) at that time. Without trees the problem of fire intensities of the past becomes a very difficult issue. Some evidence may be gleaned from requirements for seed to be shed from fire-dehiscent fruits of shrubs, from germination requirements of seeds, or from fuel accumulations and fire intensities predicted from a knowledge of past fire frequencies.

Past fire frequencies can be interpreted to some extent from a knowledge of species characteristics. Perhaps the best examples can be drawn from the presence of a fire sensitive species which reproduces from canopy seed after fire. If the primary juvenile period (Gill 1975) or time to flowering from seed is known, then the theoretical maximum frequency of fires for persistence of the species of the area is known. However, Zedler's (1977) data for *Cupressus forbesii* (family Cupressaceae) suggests that a longer period is necessary for an effective seed pool to be built up. In South Africa, Kruger (1977a) notes that fynbos species may have primary juvenile periods of up to about eight years. In Australia, a common heathland shrub *Banksia ornata* (family Proteaceae) has a primary juvenile period of five to seven years (Specht, Rayson and Jackman 1958). Another possibility is for minimum fire frequency for species persistence to be ascertained for non-sprouting species from a knowledge of their longevity. Such species have no reproductive potential from within the area once their longevity is exceeded and the store of canopy seed on the plant is lost. The presence of this type of plant, therefore, can be used to suggest a certain maximum and a certain minimum fire frequency for survival. That the survival of such a species can be threatened has been indicated for South Africa by Taylor (1977).

A less direct method of deducing past fire frequencies has recently been suggested by Keeley (1977) for chaparral communities in California. This method involved the comparison of abundance and diversity of non-sprouting and sprouting *Arctostaphylos* and *Ceanothus* (families Ericaceae and Rhamnaceae respectively). Keeley (1977) hypothesized that non-sprouting species would be promoted by less frequent fires (e.g.< 100 year intervals), while sprouting species would be promoted by more frequent fires. Support for the hypothesis came from simultaneous occurrences of high lightning-fire frequencies and sprouting species, and of low lightning-fire frequencies and non-sprouting species along elevational and latitudinal gradients in California. That large fires occurred at relatively infrequent (20-40 year) intervals in southern California in the past was suggested by Byrne, Michaelsen and Soutar's (1977) stratigraphic data.

From data such as the above, some idea of past fire regimes can be deduced. Interpretation of such data requires care, however, especially if absence of a species is used as evidence. Absence of species can be caused by many factors, but it is worth noting that a particular combination of fire events can cause the demise of a species — rather than an average fire cycle (Gill 1975). In management, data of past fire regimes can provide guidelines for restoring fire to ecosystems altered by fire suppression policies, while the species which provided the data can be used as an indicator species for management. For example, the success of a fire sensitive, reseeding species can be used as an indicator of maximum (and possibly minimum) fire frequencies for species persistence, while other species can be used to set limits to seasons of burn and intensities of fire.

Assessment of species effects on flammability

Flammability is an important topic for the land manager in a fire-prone environment. It can be important for several reasons including: the need to understand fire behaviour; the need for ecological understanding of the rôle of flammability in national park ecosystems, wildernesses and reserves; the need to manipulate fuels ecologically in some circumstances (through species composition, for example); and the need to be aware of non-flammable species which may fulfill a rôle in cultivated landscapes at the wildland-urban interface.

The study of field flammability can be difficult because of the many variables involved. In the field, the topic has been approached empirically, statistically and through principles of ignition and combustion, often derived in the laboratory (Lindenmuth and Davis 1973). In the laboratory, studies have concerned fires in cribs, weighing baskets and other apparatus, often in order to ascertain the rôles of individual variables of the field situation as they affect flammability. Many questions remain, however, and some of those relevant to the present context are proposed below.

How should flammability be measured for individual species? What is the range of variation of species flammability in a community and is this range significant? How important are live fuel components compared to dead ones? What, if any, is the adaptive significance of fuels which have low moisture, low nutrient status and high volatiles? Will an understanding of the ecology of fuel type and production (through plant growth and composition, insect and fungal susceptibility, rates of decay, etc.) make possible a wider range of options for fuel control than presently practised today?

These topics might usefully be studied on a comparative basis between countries as well as between workers within countries. Many types of studies are presently in progress; many methods are being used. Co-operation in research through individual research workers visiting internationally, through correspondence and manuscript exchange, and through appropriate scientific symposia or workshops would seem worthwhile and effective in order to help achieve answers to such problems.

Responses of invasive species to fire regimes

For the purposes of this Technical Note, 'invasive' species are considered to be those species exotic to a landscape which have spread naturally into wildlands or have been deliberately introduced for specific purposes. An example of the former would be the invasive Australian plants spreading through conservation reserves in South Africa. An example of the latter would be the herbaceous species used for erosion control on burned wildlands of California. The naturally invading exotics are a particular problem for conservation while the applied exotics might be considered in the light of their effectiveness, cost, and impact on the native plant community.

Naturally spreading invaders are a problem of considerable concern in a number of conservation areas of the world. In the mediterranean regions of South Australia, Heddle (1974) has noted the multiplication of the South African *Senecio pterophorus* DC. (family Compositae) after fire. In South Africa, the problem appears to be even more serious. Taylor (1977) notes that 'these invasive plants now present a far greater threat to our flora than the past evils of frequent burning and flower-picking combined'. Several woody weeds are involved, their origin being Europe and Australia.

The invasive woody weeds of South African reserves form thickets which suppress and eventually replace the natural fynbos of the area, which is known for its high species diversity and its large number of endemic species (Kruger 1977b, Taylor 1977). Despite efforts to control the invaders, these exotic species have increased 'alarmingly' in the last ten years (Taylor 1977). How could a knowledge of adaptations to fires be used to help control these species? Clearly, the species are well adapted to the presence of fires in Cape fynbos today and show a number of typical adaptive traits to fires. The *Acacia* species (family Leguminosae) have long-lived seed and *Hakea's* (family Proteaceae) have hard woody fruits which are stimulated to release their seed by fire occurrence. What options seem likely to provide methods of control?

Taylor (1977) has already noted that fires may have a rôle in the control of *Acacia* through control (by spraying) of seedlings emergent after fire. Some control may also be possible through control of seed germination by fire intensity. For *Hakea* the strategies adopted could be different. The vegetative plants are apparently sensitive to fire but release seed from their hard woody fruits whenever a fire occurs. There is no seed storage in the soil and the key to control may be found in the knowledge of the ecology of the hard woody fruit in relation to fire and parasitism.

The hard woody fruits of species such as *Hakea* are stored on the plant and build up in numbers until opened by fire. In their natural locations in Australia, some losses occur to this seed bank through the depredations of insect predators. Some interesting research on this aspect of the ecology of the hard woody fruits has already been carried out (but not yet published) and could provide the key to biological control of the species in southern Africa. Co-operative research between southern Africa and Australia on this subject would seem most worthwhile.

Other invasive species of exotic origin occur in other mediterranean regions although documentation of these seems to be lacking. Communication between relevant authorities in each mediterranean-climate country may enable a better understanding of the scope of the problem, on the one hand, and a start to research on likely means of control, on the other.

Productivity of shrublands under grazing and burning regimes

Shrublands in mediterranean lands are used for the production of domestic and game meats. In many circumstances, the effects of management on plant species' survival, growth and reproduction is unknown, yet is important to long-term productivity. In some cases grazing may eliminate the option to burn. What rôle does species composition have in these systems? How do the individual species which form the communities respond to different grazing and fire regimes?

Browse is a particularly important food source for deer and some other wildlife in North America and South Africa. Knowledge of the sprouting adaptation to fires may enable more successful management for favoured browse species. In some areas such as the South Eastern United States, annual burning may cause a decline in some sprouting populations (Grano 1970) while an absence of burning may allow trees and shrubs to grow beyond the reach of animals or seriously decline in numbers and palatability. Some optimal fire regime for this purpose could be devised. Cable (1975) has called for 'refined prescriptions for using fire as a tool for controlling sprout growth or maintaining it in a young, nutritious condition through timing and frequency of reburning'. In South Africa, on the other hand, injudicious grazing with frequent burning of fynbos lands often leads to development of monospecific stands of unpalatable renosterbos, *Elytropappus rhinocerotis* Less (family Compositae). While the ecology of this species is not fully understood, studies on reproductive biology have, for example, contributed toward the understanding which can permit successful control of the shrub through appropriate vegetation and stock management (Levyns 1956).

In European and North African lands, grazing by domestic animals such as goats is an important industry. Little appears to be known of the rôles of burning and grazing on productivity and species responses. Further study might be found to be worthwhile after a review of the problem. Such a topic might well be studied through a pilot project in the Mediterranean Sea region, perhaps within the framework of co-operative activities being developed among mediterranean countries under the Man and the Biosphere (MAB) Programme.

BIBLIOGRAPHY

BYRNE, R.; MICHAELSEN, J.; SOUTAR, A. 1977. Fossil charcoal as a measure of wildfire frequency in southern California: a preliminary analysis. In: *Proceedings of the Symposium on the Environmental Consequences of Fire and Fuel Management in Mediterranean Ecosystems.* H.A. Mooney and C.E. Conrad (Eds), p. 361-367. USDA Forest Service Gen. Tech. Rep. WO-3. US Department of Agriculture, Washington, D.C.

CABLE, D.R. 1975. *Range management in the chaparral type and its ecological basis: the status of our knowledge.* USDA For. Serv. Res. Pap. RM-155. US Department of Agriculture, Washington, D.C.

GALBRAITH, A.V. 1937. *Mountain ash (Eucalyptus regnans - F. Muell.). A general treatise on its silviculture, management and utilization.* Govt. Printer, Melbourne.

GILBERT, J.M. 1959. Forest succession in the Florentine Valley, Tasmania. *Proc. Roy. Soc. Tas.*, 93, p. 129-151.

GILL, A.M. 1975. Fire and the Australian flora: a review. *Aust. For.*, 38, p. 4-25.

GILL, A.M. 1977a. Management of fire-prone vegetation for plant species conservation in Australia. *Search*, 8, p. 20-26.

GILL, A.M. 1977b. Plant traits adaptive to fires in Mediterranean land ecosystems. In: *Proceedings of the Symposium on the Environmental Consequences of Fire and Fuel Management in Mediterranean Ecosystems.* H.A. Mooney and C.E. Conrad (Eds), p. 17-26. USDA For. Serv. Gen. Tech. Rpt. WO-3. US Department of Agriculture, Washington, D.C.

GIMINGHAM, C.H. 1971. British heathland ecosystems: the outcome of many years of management by fire. *Tall Timbers Fire Ecol. Conf. Proc.*, 10, p. 293-321.

GRANO, C.X. 1970. *Eradicating understory hardwoods by repeated prescribed burning.* USDA For. Serv. Res. Paper SO-56. Southern Forest Experiment Station.

HEDDLE, E.M. 1974. South African daisy in the National Parks of South Australia. *Environ. Conserv.*, 1, p. 152.

HOUSTON, D.B. 1971. Research on ungulates in northern Yellowstone National Park. In: *Research in National Parks.* AAAS Symposium. USDI National Park Serv. Symp. Series 1. US Department of the Interior, Washington, D.C.

JACOBS, M.R. 1955. *Growth Habits of the Eucalypts.* Govt. Printer, Canberra.

KAYLL, A.J. 1974. Use of fire in land management. In: *Fire and Ecosystems.* T.T. Kozlowski and C.E. Ahlgren (Eds), p. 483-511. Academic Press, New York.

KEELEY, J.E. 1977. Fire-dependent reproductive strategies in *Arctostaphylos* and *Ceanothus*. In: *Proceedings of the Symposium on the Environmental Consequences of Fire and Fuel Management in Mediterranean Ecosystems.* H.A. Mooney and C.E. Conrad (Eds), p. 391-396. USDA For. Serv. Gen. Tech. Rpt. WO-3. US Department of Agriculture, Washington, D.C.

KESSELL, S.R. 1976. Wildland inventories and fire modelling by gradient analysis in Glacier National Park. *Tall Timbers Fire Ecol. Conf. Proc.*, 14, p. 115-162.

KILGORE, B.M. 1973. Impact of prescribed burning on a sequoia-mixed conifer forest. *Tall Timbers Fire Ecol. Conf. Proc.*, 12, p. 345-375.

KRUGER, F.S. 1977a. Ecology of Cape fynbos in relation to fire. In: *Proceedings of the Symposium on the Environmental Consequences of Fire and Fuel Management in Mediterranean Ecosystems.* H.A. Mooney and C.E. Conrad (Eds), p. 230-244. USDA For. Serv. Gen. Tech. Rep. WO-3. US Department of Agriculture, Washington, D.C.

KRUGER, F.S. 1977b. Ecological reserves in the Cape fynbos: towards a strategy for conservation. *A. Afr. J. Sci.*, 73, p. 81-85.

LEVYNS, M.R. 1956. Notes on the biology and distribution of the rhenoster bush. *S. Afr. J. Sci.*, 52, p. 141-143.

LINDENMUTH, A.W.; DAVIS, J.R. 1973. *Predicting fire spread in arizona's oak chaparral.* USDA For. Serv. Res. Pap. RM-101. Rocky Mountain Forest and Range Experimental Station, Fort Collins.

NOBLE, I.; SLATYER, R.O. 1977. Post-fire succession of plants in mediterranean ecosystems. In: *Proceedings of the Symposium on the Environmental Consequences of Fire and Fuel Management in Mediterranean Ecosystems.* H.A. Mooney and C.E. Conrad (Eds), p. 27-36. USDA For. Serv. Gen. Tech. Rep. WO-3. US Department of Agriculture, Washington, D.C.

SPECHT, R.L.; RAYSON, P.; JACKMAN, M.E. 1958. Dark Island Heath (Ninety-Mile Plain, South Australia) VI. Pyric succession: changes in composition, coverage, dry weight and mineral nutrient status. *Aust. J. Bot.*, 6, p. 59-88.

TAYLOR, H.C. 1977. Aspects of the ecology of the Cape of Good Hope Reserve in relation to fire and conservation. *Proceedings of the Symposium of the Environmental Consequences of Fire and Fuel Management in Mediterranean Ecosystems.* H.A. Mooney and C.E. Conrad (Eds), p. 483-487. USDA For. Serv. Gen. Tech. Rep. WO-3. US Department of Agriculture, Washington, D.C.

WAGTENDONK, J.W. VAN. 1974. *Refined burning prescriptions for Yosemite National Park.* USDI National Park Serv. Occ. Pap. 2. US Department of the Interior, Washington, D.C.

ZEDLER, P.H. 1977. Life history attributes of plants and the fire cycle: a case study in chaparral dominated by *Cupressus forbesii*. *Proceedings of the Symposium on the Environmental Consequences of Fire and Fire Management in Mediterranean Ecosystems.* H.A. Mooney and C.E. Conrad (Eds), p. 451-458. USDA For. Serv. Gen. Tech. Rep. WO-3. US Department of Agriculture, Washington, D.C.

Ecological impact of fires on mineral and sediment pools and fluxes

Philip W. Rundel

BACKGROUND

While mediterranean-climate regions of the world are generally considered as low nutrient ecosystems, distinct differences exist in the nutrient regimes of these five areas. In contrast to the typically convergent pattern of ecosystem characteristics of mediterranean-climate regions, soil nutrient levels exhibit remarkably divergent conditions. With respect to two primary nutrients, nitrogen and phosphorus, these differences provide a natural experiment in studying the adaptations of plants to low nutrient conditions. Related to this is the rôle of natural fire in influencing nutrient cycles and the means by which fire-adapted plant species can utilize fire-nutrient cycling. With increasing human perturbations of natural fire frequencies and intensities, it is vitally important that the impact of fire on nutrient cycling and the ecological consequences of past and potential future human activities in these patterns be understood.

Two mediterranean-climate regions – Chile and the mediterranean zone of Europe, North Africa and the Middle East – are characterized by moderately fertile soils (Fig. 1). Both nitrogen and phosphorus occur in moderate levels, with pH relatively close to neutral (or alkaline in many calcareous soils of southern Europe) to promote high availability of the latter. Soil nutrient levels are not strongly limiting to productivity over broad areas. Contrasted to this condition are the extremely infertile soils of Australia (Groves 1978) and South Africa. These soils, with unusually low levels of phosphorus and nitrogen, have developed from geologically very old parent materials (typically quartz or sandstone) which are heavily weathered and often podsolized. Soil nutrients are too low to produce any significant growth of agricultural plants. The typical acidic nature of the soil solution makes even the meager amounts of total phosphorus relatively unavailable, and thus plant productivity is phosphorus-limited. Growth of dominant sclerophyll species increases two to three times with the addition of phosphate fertilizer (Specht 1963; Specht and Groves 1966; Heddle and Specht 1975). California represents an intermediate condition of soil nutrient levels. Soils are moderately low in both nitrogen and phosphorus, but not to the extreme levels of Australia and South Africa. Fertilization experiments indicate that productivity is strongly nitrogen-limited (Hellmers, Bonner and Kelleher 1955).

Although nutrient studies have been carried out in many mediterranean-climate regions, no extensive baseline data exist on natural patterns of nutrient cycling which include information on seasonal changes in nutrient pool size and on the magnitude of fluxes of nutrients between pools. The impact of fire on these cycles and its significance in promoting long term stability has not been investigated. Isolated studies have identified individual species adaptations to increase uptake, storage and efficient utilization of nutrients, but broad investigation on generalized adaptations in this regard have not been carried out. Participants at the Stanford workshop considered international co-operation essential to provide the kinds of comparative data required to develop an understanding of primary nutrient cycling patterns. In addition, they felt that it is extremely important to develop guidelines to standardize methods of sample collection and analysis, as much existing data is not comparable due to lack of such standardization.

RESEARCH PRIORITIES

At the Stanford workshop, the participants agreed on five research priorities that are of primary importance in understanding the ecological impact of fires on mineral and sediment pools and fluxes in mediterranean ecosystems.

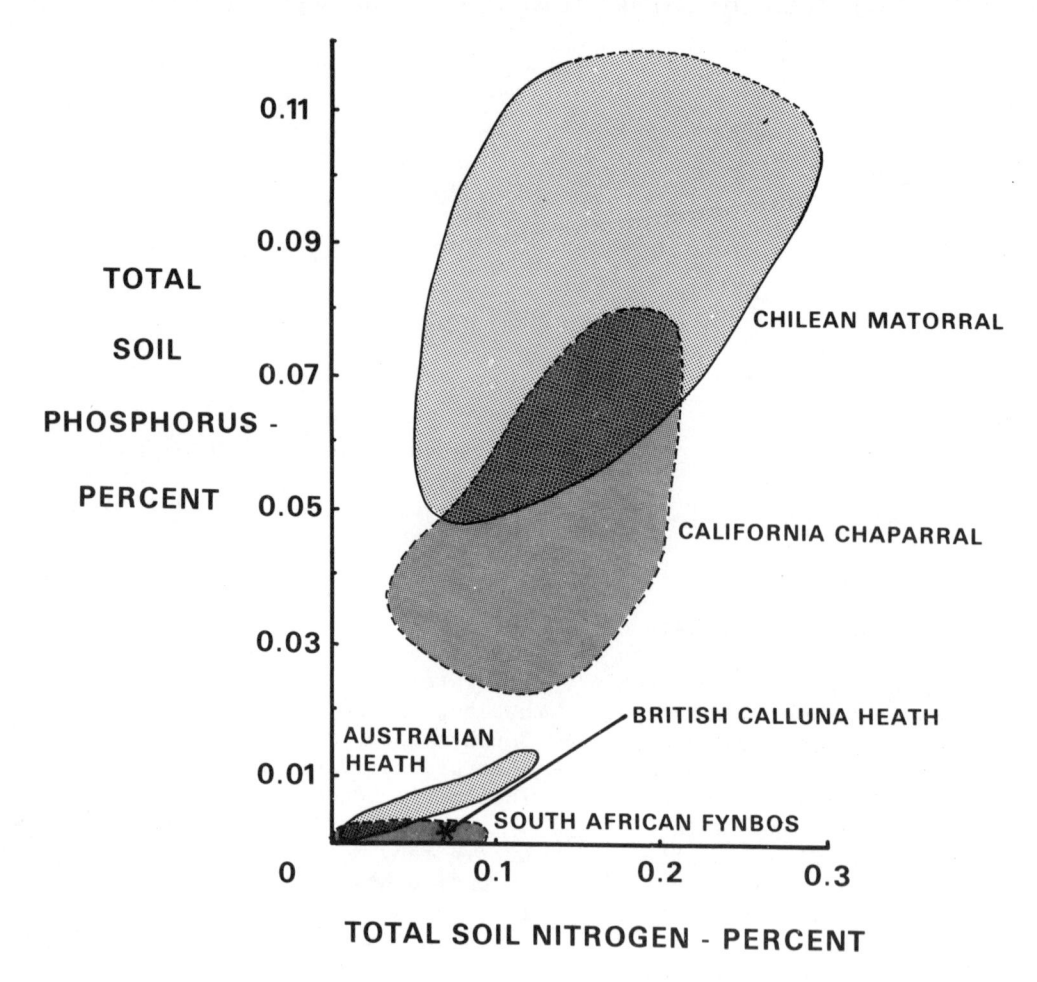

Figure 1. *Soil fertility in mediterranean-climate regions as illustrated by levels of phosphorus and nitrogen.*

These five areas each represent considerable gaps in our understanding of these important processes and provide a clear example of the necessity of co-operative international studies in gaining this understanding.

Nitrogen and phosphorus pool sizes

Although a variety of nutrients including cations, sulfur, iron, and selenium may locally limit growth of mediterranean-climate vegetation, nitrogen and phosphorus are the nutrients which most commonly limit plant productivity. While individual studies of local areas have looked at some aspects of plant, litter and soil compartments sizes for these nutrients, no complete data have been published for any single area. Comparative studies of seasonal changes in total and available forms of nitrogen and phosphorus for important ecosystem compartments would provide an important means of assessing the relative nutrient deficiency for each of the five mediterranean-climate regions of the world.

As a first stage in producing this type of comparative study, a sample programme should be developed as outlined in Table 1. This programme would include ten geographic sites, two nutrients (plus available forms for soil), four seasons, five ecosystem compartments, and two fire successional stages. Geographical sites for study could include three areas in California where such studies are already in progress, two Australian community types, Israel, France, Greece, and South Africa. Data for total nitrogen and total phosphorus should be collected for mature leaves, mature stems, roots, fresh litter and mineral soil (0-10 cm). In addition, NH_4, NO_2 and soluble phosphorus data should be collected for the latter compartment. Four seasonal samples are projected: initiation of primary growing season, mid-growing season, end of primary growing season, and middle of drought period. Two fire successional stages should be

Table 1. *Sample design for pilot study of nitrogen and phosphorus compartment pool sizes.*

Nutrients	Compartment	Season	Successional Stage	Geographic Location	
Total N	Leaves-mature	Initiation of growing season	Mature stand	California	- San Diego
Total P	Stems	Mid-growing season			- San Dimas
	Roots	End of growing season	Post fire		- Sequoia National Park
	Litter	Autumn		Australia	- heath
	Soil (0-10 cm)				- Mallee
				Chile	
				Israel	
				France	
				Greece	
				South Africa	

considered: mature stands older than twenty years and post-fire stands in their first year since burning. These later data are particularly critical to assess the importance of fire mineralization of nutrients in natural communities.

Nitrogen and phosphorus fluxes

While complete comparative data on all aspects of nutrient cycling patterns of mediterranean-climate regions and their relationship to fire interactions must remain a long-term goal, comparative data on the magnitude and significance of important nitrogen and phosphorus fluxes must remain an important priority. Of particular interest are questions relating to fluxes promoting more efficient cycling of nitrogen in low nitrogen environments or similarly efficient phosphorus cycling in low phosphorus environments. Under the broad heading of comparative studies of fluxes, a variety of specific projects are of high priority:
- rôle of microbial activity in influencing nitrogen and phosphorus fluxes;
- significance of nitrogen and phosphorus fluxes as limiting factors for primary productivity. Carefully designed fertilization experiments under both field and greenhouse conditions are recommended here;
- significance of nitrogen fixation, with consideration of legumes, non-legume nodulated taxa, and freeliving microorganisms;
- decomposition rates of litter, with consideration of environmental as well as extrinsic biological variables. Particular consideration should be given to rates of nitrate release;
- magnitude and importance of litter fall in relation to nutrient deficiency and physiological stress;
- variation of phosphate release and immobilization under natural environmental conditions;
- significance of atmospheric inputs of nitrogen and phosphorus under both natural conditions and semi-natural conditions with industrialization.

Systems models of nutrient cycling

Future development of enlightened fire management policies for mediterranean-climate vegetation regions throughout the world requires an ability to accurately predict the consequences of varying types of natural and management fires on patterns of nutrient cycling. This goal can best be achieved by developing a set of models for nutrient systems to provide reliable estimates of the consequences of individual management decisions or natural perturbations on existing patterns. These models must be based on quantitative understanding of the essential process of nutrient flow through these ecological systems and of the impact of fire frequency and intensity on these processes. Existing programmes of research on fire-nutrient interactions in mediterranean-climate areas of Australia and California provide a data base from which such models can be developed, but a gap in our understanding of the magnitude of many significant nutrient fluxes must be filled before operational simulations can be made.

Adaptations to utilize fire-nutrient cycling

It is known that fire is able to break down forest organic matter and leave at least a portion of its nutrient content in the soil in available forms (Hatch 1960; Christensen and Muller 1975; St. John and Rundel 1976). Since many species have evolved dependence of periodic fires for regeneration and other aspects of their existence (Kowzlowski and Ahlgren 1974), it is reasonable to hypothesize that some of these species have also evolved a dependence on fire as their natural mineralizing agent. With fire as the primary decomposer, fire-adapted species would be at a competitive advantage in nutrient poor soils; minerals would be present in available form only at a time when non-fire-adapted species would be absent from the environment. Thus it is hypothesized that fire-adapted species which survive relatively frequent ground fires under natural conditions exhibit a syndrome of specializations that cause the nutrients in their lost foliage to be preferentially released by fire rather than biological decomposers, and that allow the fire-adapted species to take best advantage of the sudden release of nutrients when a fire occurs.

Field observations and data in the literature suggest that this set of specializations may in fact exist. Leaf litter of many fire-adapted species is slow to decompose, often because of chemical composition (Mount 1964), leaf geometry that encourages rapid drying (Vogl 1974), or other factors that are under genetic control. After a fire, it is common for fire-adapted species to increase growth rate (Hatch 1960; Cremer 1962; Pearson, Davis and Schubert 1972), increase foliar nutrient content (Lay 1957; Steward and Ornes 1975), or increase production of flowers or fruits (Daubenmire 1968; Vogl 1968). All of these suggest an increased uptake of mineral nutrients. These observations are merely suggestive of the idea that the plants depend on fire as a mineralizing agent. An experimental test requires a defined set of species for comparison and a series of testable hypotheses upon which the argument can be based.

If fire-adapted plant species possess an evolutionary dependence on the action of natural fires to maintain efficient patterns of nutrient cycling, it should be possible to demonstrate the qualitative and quantitative characteristics of the plants. Both morphological and physiological manifestations of this adaptation can by hypothesized.

Geomorphological changes

Few data are available to assess the significance of fire in promoting erosion and other geomorphological changes in sediment pools. The rôle of human activities in altering natural fire frequency and intensity must have a very important influence on such changes, but documentation for such effects is needed. This phenomenon may be much more important in specific regions than others. The effects are often severe, for example, in the mountains of southern California.

BIBLIOGRAPHY

CHRISTENSEN, N.; MULLER, C.H. 1975. Effects of fire on factors controlling plant growth in *Adenostoma* chaparral. *Ecol. Monog.*, 45, p. 29-55.

CREMER, F.W. 1962. Effects of burnt soil on the growth rate of eucalypt seedlings. *Inst. Foresters of Australia, Newsletter*, 3 (3), p. 2-5.

DAUBENMIRE, R. 1968. Ecology of fire in grasslands. *Adv. Ecol. Res.*, 5, p. 209-266.

GROVES, R.H. 1978. Nutrient cycling. In: *Heathlands and related shrublands*. R.L. Specht (Ed.). Elsevier Publ., Amsterdam. (in press).

HATCH, A.B. 1960. *Ash bed effects in western Australian forest soils*. Bull. For. Dept. West. Australia 64.

HEDDLE, E.M.; SPECHT, R.L. 1975. Dark Island heath (Ninety-mile Plain, South Australia) VIII. The effects of fertilizers on composition and growth, 1950-1972. *Aust. J. Bot.*, 23, p. 151-164.

HELLMERS, H.; BONNER, J.F.; KELLEHER, J.M. 1955. Soil fertility: a watershed management problem in the San Gabriel Mountains of southern California. *Soil Sci.*, 80, p. 189-197.

KOWZLOWSKI, T.T.; AHLGREN, C.E. (Eds) 1974. *Fire and ecosystems*. Academic Press, New York.

LAY, D.W. 1957. Browse quality and the effects of prescribed burning in southern pine forests. *Jour. For.*, 55, p. 342-349.

PEARSON, H.A.; DAVIS, J.R.; SCHUBERT, G.H. 1972. Effects of wildfire on timber and forage production in Arizona. *J. Range Manage.*, 25, p. 250-253.

ST. JOHN, T.V.; RUNDEL, P.W. 1976. The rôle of fire as a mineralizing agent in a Sierran coniferous forest. *Oecologia (Berlin)*, 25, p. 35-45.

SPECHT, R.L. 1963. Dark Island heath (Ninety-mile Plain, South Australia). VII. The effect of fertilizers on composition and growth, 1950-1960. *Aust. J. Bot.*, 14, p. 201-221.

SPECHT, R.L.; GROVES, R.H. 1966. A comparison of the phosphorus nutrition of Australian heath plants and introduced economic plants. *Aust. J. Bot.*, 14, p. 201-221.

STEWARD, K.K.; ORNES, W.H. 1975. The autecology of sawgrass in the Florida everglades. *Ecology*, 56, p. 162-171.

VOGL, R.J. 1968. Fire adaptations of some southern California plants. *Proc. Tall Timbers Fire Ecol. Conf.*, 7, p. 79-109.

VOGL, R.J. 1974. Effect of fire on grasslands. In: *Fire and ecosystems*. T.T. Kowzlowski and C.E. Ahlgren (Eds), p. 139-194. Academic Press, New York.

Ecological impact of fires on community productivity, structure and diversity

John W. Menke

INTRODUCTION

Virtually all mediterranean-climate ecosystems evolved with fires as a major abiotic influence on vegetation community production, structure and diversity. Human alteration of fire impacts has resulted in a broad array of ecosystem states that can be grouped into three categories with respect to fire:
- minimum human impact ecosystems;
- moderately altered ecosystems;
- highly altered ecosystems.

Societies in mediterranean-climate regions of the world have set different land use and land management goals for each of these ecosystems. Human perception of the devastating nature of fire has influenced the rôle fire has played in these fire dependent ecosystems. Providing the land manager with more knowledge on the ecological impact of fires would allow more goal oriented decisions.

Starting with the most threatened ecosystems, minimum human impact ecosystems are those perceived to have been the least altered by man and to possess the attributes of natural mediterranean-climate ecosystems. It is unlikely that any mediterranean-climate ecosystem is totally unaltered, but many of the ecosystems in the first group are considered natural. Nature conservation is the management goal for these ecosystems. The productivity, structure and diversity of plant and animal communities undergoing natural fire impacts provide a baseline for comparison with ecosystems with altered fire inputs.

Moderately altered ecosystems can be defined as those systems that are incurring altered fire inputs such as more or less frequent burning than naturally occurred, different time series distribution of fire intensities, or different seasons of burning. In some moderately altered ecosystems, larger numbers of herbivores now consume some of the biomass that was naturally consumed by fire in the past. Generally these ecosystems have a lower food and fiber production capability than highly altered ecosystems and are extensively, not intensively, managed.

Finally, highly altered ecosystems are those systems that man typically manages for export of food in the form of meat from grazing animals and fiber in the form of wood products. In the former case, fire can be used only periodically to alter species composition to favour desirable forage plants. In the latter case, man has attempted to exclude fire completely to protect his fiber products or has dramatically altered the occurrence and severity of fire. All too often the frequency of fire is successfully reduced but the increase in intensity and size of fires may cancel the effect of lowered frequency. Even with highly developed technology, it is extremely difficult, if not impossible, to keep fire out of any of the three ecosystem categories.

RESEARCH RECOMMENDATIONS

Participants at the Stanford workshop suggested that long term demonstration and research plots be selected and established for the minimum, moderately altered, and highly altered ecosystems in all mediterranean-climate regions of the world. There is an immediate need for the identification of rare plant and animal species and associations. In all cases emphasis should be on holistic, multidisciplinary studies of flora and fauna. Because man has had an impact on natural fire regimes in virtually all areas, it is imperative that analyses take place as soon as possible to document past vegetation structures and fire histories, if possible, for the variously altered ecosystems. Studying the landscape today and attempting to project the vegetation and animal populations backward in time would provide our best estimate of pristine community structure.

Diversity

Beyond this first recommendation the relative availability of research funding greatly influenced the kinds of research recommended. It was generally agreed that diversity data were the minimum information required of all research. Those investigators having only minimum funding should concentrate on plant and animal species diversity and on the impact of at least the above outlined three levels of fire input. One major void recognized at this level of research is the almost complete lack of insect data. It was also pointed out that in the California mediterranean-climate ecosystem little data exist even at the diversity level for various levels of fire input.

Structure

At the next level of research, that is research dealing with the spatial structure of vegetation and animal populations, it was agreed that joint studies involving plants and animals should be conducted together. Four basic questions were proposed. First, what is the rôle of fire in the maintenance or improvement of ecosystems? Second, what is the rôle of introduced fire? Third, what is the rôle of fire in exotic components of ecosystems? Fourth, what is the rôle of fire in maintenance of altered communities? Each of these questions takes on a different connotation for the three variously altered mediterranean-climate ecosystems.

The animal component can be divided into two parts:
- What are the habitat requirements and the impact of vegetation structure and diversity on native fauna?
- What are the requirements or how might one enhance the space, food and cover requirements of purposefully introduced exotics such as livestock and game animals?

Intensive studies of vegetation have recently proved very useful in explaining animal population structure. This means that it should be possible to use detailed studies on fire behaviour in various vegetation complexes to develop biologically meaningful models characterizing animal habitats.

Two related approaches have emerged concerning community structure. First, a morphological and ecophysiological approach attempts to characterize the above and belowground structure of mediterranean-climate ecosystems: the aboveground component being mainly light determined, and the belowground component being soil nutrient and water determined. A second approach has been more directly fire related and has sought to characterize the aboveground structure only from a fine fuel availability and fire hazard point of view. This approach is ecophysiological only to the point that fuel moisture content of live biomass is a physiologically determined attribute of the plant. It is this second approach that some of the workshop participants feel is imperative to understand structure and structural dynamics of mediterranean-climate ecosystems.

Vegetation structure description in fire dependent ecosystems must include estimates of available fuel (where available fuel is defined as that portion, i.e. weight, of live and dead biomass that contributes energy to the passing fire front). Only by knowing the available fuel at the time of burning can one estimate the fire intensity and thus the impact of fire on post-fire community structure. Parameters of aboveground structure necessary to estimate available fuel are dry weights of plant parts by species, the fuelbed configuration, the fuel composition including ash content, moisture content, flammable organic chemical compound content, and the spatial mosaic and layering pattern in the community. Given those parameters, models need to be developed that describe fire intensity, fuel consumption and general community structure dynamics as a vegetation succession process.

Ecophysiological and morphological approaches to the study of community structure remain quite appropriate. However, spatial and temporal structural dynamics of mediterranean-climate ecosystems must be described to determine the impact of natural or altered fire input. In both cases, short term (daily or weekly) weather data will be required to drive such models of fuel dynamics. Topographic relief will act to significantly modify local weather conditions and must be taken into account. This latter point has been the most serious failure of models developed thus far. Land managers experienced with fire have emphasized the importance of local site conditions on fire behaviour.

More interesting hypotheses can be formulated by taking an explanatory systems modelling approach or even a non-mathematical conceptual modelling approach. For example, what is the effect of vegetation gradients in space and over time on fire impacts and faunal responses? The complexity of a process as complicated as fire requires a computer modelling approach and thus more research in systems models of fuel dynamics is recommended.

Productivity

Finally, research in the area of productivity of mediterranean-climate ecosystems, depending on the level of resolution one is interested in pursuing, may include the disciplines of forestry, range management and agronomy. Ecological productivity was the emphasis stressed by the Stanford workshop. Basic questions asked included, for example: Can fire-adapted mediterranean scrub maintain its productivity under altered fire inputs? Would nutrients be adequately conserved under more frequent burning programmes suggested by fuel managers? What is the impact of less frequent, more intense, larger wildfires on short and long term community productivity?

The nature of productivity studies requires considerable effort and funding, especially if they are to be interdisciplinary. Workshop participants felt that collaborative efforts should be mounted in the Mediterranean Basin, Australia, the United States and in South Africa and that major efforts should be made to coordinate the methodological approaches and research priorities.

Alternative fire-management policies: methods for assessment

James K. Agee

This chapter concentrates on two levels of fire research definition for alternative fire management policies. First, there must be a series of analytical models sufficient to provide a conceptual framework for research. Second, there must be identification of factual research needs to fill the gaps in the framework. With such information, managers can make informed decisions about what strategy or combination of strategies is most likely to contribute to a fire management policy consistent with their land management objectives.

A basic problem is that of defining what is meant by an 'alternative fire management policy'. A fire management policy is comprised of one or more fire strategies or actions designed to achieve land management objectives. Fire management strategies can include various forms of fire use and control (or non-control) as well as various forms of fuel management without fire, such as mechanical treatment or vegetation type conversion.

The strategies that comprise most fire policies today include:
- fire suppression (with or without alternative measures);
- prescribed fire;
- natural fire;
- no management at all.

Fire management policies consist of various combinations of these strategies but 'no management' is becoming rare in this age of scarce resources. Although there are over a dozen potential policies, the most common in practice are:
- total fire suppression;
- fire suppression with alternative fuel treatment;
- prescribed fire with fire suppression;
- natural fire with fire suppression;
- fire suppression, prescribed fire, and natural fire.

The final goal of any fire management strategy is to achieve overall land management objectives. There is, therefore, no single combination of strategies that will achieve all land management objectives for all managers; moreover, a manager who has found a satisfactory combination, may well find it inapplicable in the future due to changing states of the biosocial system in which he or she operates.

The existing literature on impacts of fire management policies is too voluminous to be summarized in this Technical Note. Agee (1974) summarizes the environmental impacts of alternative fire management policies.

POLICY MODELS

Rational decisions on fire management policies are based on two levels of analysis: knowledge of each alternative policy, and knowledge of the relative effects of each policy. The former will be discussed in the latter portions of this chapter. The second level of analysis requires some method of weighting the natural and cultural factors of each policy alternative. There is thus a conceptual problem involved in attempting to develop adequate analytical models that will help formulate fire policy. The conceptual problem must be dealt with before factual ones, for it will determine how the factual data will be collected.

A number of analytical models are already available that would benefit from improved factual data and conceptual approaches. Any good model is a beginning, a framework around which a better model can be developed. The FOCUS model (or Fire Operational Characteristics Using Simulation) developed at the Riverside Fire Laboratory was designed to quickly evaluate alternative fire management plans (Storey 1972). It considers weather, fuels, topography, suppression forces, and the transportation network to plan for an overall fire control organization. The cost effectiveness of changes in ground crews, air support, hazard reduction programmes, or fire-break location can be compared and ranked. While FOCUS is not a 'real-time' system, it can contribute to a wide range of fire policy decisions.

Another recent model is the gradient modelling approach as applied to Glacier National Park (Montana, U.S.A.) and now being applied to southern California chaparral (Kessell 1976). This type of system is oriented to natural resources inventory with updating, and simulations and predictions of natural processes and management actions.

Computer modelling alone cannot develop all of the information needed for fire policy development. There are obvious constraints such as slowly changing cultural factors or legislation in various countries or states that raise 'red flags' at certain actions. Furthermore, on many lands, public involvement by way of environmental assessments or statements may precede and alter policy implementation. However, systems analysis is sure to become an increasingly useful and important tool of the fire planner. Acceptance or alteration of existing models, or development of new models, is the most logical way to identify and carry out new research needs for a particular management unit. It is suggested that the research projects developed in other chapters of this Technical Note be placed in a systems context. This recommendation applies to all research projects and is the best way both to avoid piecemeal projects that have limited applicability and to provide continuity and co-ordination between large research projects.

FIRE SUPPRESSION RESEARCH

Historically, fire suppression has been widespread in most mediterranean-climate ecosystems. While the effectiveness and efficiency of fire control operations has increased over time, the number of large conflagration fires has not dropped. This is due to the ineffectiveness of present fire suppression methods and equipment in stopping the head of hot wildfires in strong winds. New firefighting techniques and equipment are not likely to change the situation, although on non-conflagration fires such advances may be effective.

Strong fire control capability will always be an important part of fire management in mediterranean ecosystems, but highest research priority should be assigned to fuel modification. The research priorities identified for impacts of fire suppression include: fuel modification, maintenance of modified fuel patterns, tools and techniques for suppression, and ecological effects of suppression policy.

Fuel modification

Research in fuel modification appears to offer the best way to increase suppression capability. The evolution of this approach has progressed from the fire-break concept through the fuel-break concept to the fuel mosaic concept (Countryman 1974). Breaking up large, uniform fuels can be done in many ways, each with its ecological and socio-economic impacts. Techniques can include mechanical and chemical treatment, prescribed fire, or modified suppression to obtain an age-class mosaic of fuel, with fuel-break and type conversion as appropriate (Philpot 1974). Research is needed to define the effectiveness of fuel modification in fire suppression and to assess the impacts of various techniques of modification.

Maintenance of modified fuel patterns

Once the desired fuel pattern is obtained, maintenance of the pattern is essential. Maintenance may be achieved using the same techniques that created the pattern, or other techniques (such as grazing) might be employed with less impact. Solving this problem involves knowledge of fuel dynamics as well as relative impacts of maintenance techniques.

Tools and techniques for suppression

Research in this area should concentrate on maximizing effectiveness and minimizing adverse impacts. Effectiveness can be increased in developing new tools for the individual firefighter, and by continuing research in fire retardant chemical properties, effectiveness and delivery systems. Strengthening the flow of information during ongoing fires, by techniques such as infrared imagery or improved fire behaviour information would also aid fire suppression effectiveness.

Minimizing adverse impacts of suppression efforts is also important. Temporary roads that later create slope stability problems, and fish kills by fire retardants in streams, are types of impacts that can be avoided or minimized through application of research findings.

Ecological effects of suppression policy

A suppression policy covers a wide range of ecological effects, ranging from the effects of successful fire exclusion to the effects of disastrous wildfires. These effects which vary considerably are essentially a baseline against which the effects of alternative policies must be assessed. In the past, there has been considerable documentation of wildfire effects, and major emphasis should be placed on arranging existing research, identifying factual gaps, and filling those gaps. This topic is dealt with elsewhere in this Technical Note. There are few areas where large-scale integrated programmes exist to evaluate alternative fire suppression techniques. Two of the most recent projects are the Laguna-Morena Demonstration Area in California and the Des Maures area in France.

PRESCRIBED FIRE RESEARCH

The ecological impacts of prescribed fires depend not only on site variables but on the kind of burn. Prescribed fire has tremendous flexibility. It can be low or high intensity, a pile burn or a broadcast burn, a heading, flanking, or backing fire; it can be the only treatment applied or only one step in a series of manipulations. The focus of future research should be to quantify the effects of such fires in terms transferable to managers. Researchers must not only provide information on soil, vegetation, or hydrologic effects but also tie these effects to specific burning conditions - in short, provide a description of the prescription.

The best system being used at present in the United States to describe fire behaviour for management is the National Fire-Danger Rating System. This system is becoming the basis for predicting prescribed fire behaviour on most public lands. Based on one of several fuel models, a Spread and Energy Release Component is calculated using inputs of wind, slope, and fuel moisture. These components are combined into a Burning Index to describe relative fire behaviour. This information, combined with other variables such as ignition pattern and aspect, provides the manager with a means of predicting fire behaviour. If research results are also tied to the National Fire-Danger Rating System, a rational analysis of effects of prescribed fires is possible. There is a lesser need for new research than there is to apply what is already known and to co-ordinate future research - this is by far the most glaring deficiency of the present literature. Other countries with fire-danger rating systems could also consider the need for co-ordinating fire effects with a standardized fire-danger rating system. The research priorities for prescribed fire can be summarized in several categories as: refinement of burning prescriptions, improved short-term weather prediction, fuel dynamics, economic studies and refined smoke management systems.

Refinement of burning prescriptions

A burning prescription is simply a set of conditions under which a fire is ignited to achieve specific objectives. This set may include fireline specifications, staffing needs, weather parameters or indexes, method of firing, and other chosen constraints. Firing methods and weather have the most direct bearing on ecological impact for given site conditions. Quantifying fire effects under varying weather and firing prescriptions increases predictability of fire use, at present perhaps the greatest obstacle to increased application of prescribed fire. It would also provide better information on controllable limits to fire in various mediterranean-climate ecosystems and the ecological effects associated with various levels of fire intensity. Training of prescribed fire managers is also an important factor.

Improved short-term weather prediction

Mediterranean climates are subject to rather rapid weather changes which can markedly influence fire behaviour. In summer, clear skies, dry air, and a nearly vertical sun provide ideal conditions for daytime heating, and such conditions are also conducive to rapid nocturnal cooling. Along coastal areas, summer fog may prevail. Sea breeze and foehn wind effects interact with local thermal winds. All of these factors may rapidly alter actual fire behaviour from what would have been predicted a few hours earlier. Prescribed fire can be applied in every season of the year, so that improvement of weather prediction must consider more than just dry-season variation. An adequate density of automated fire weather monitoring, and access to it through a system such as AFFIRMS (a United States Forest Service computer system for calculating fire-danger rating), would provide better information for fire managers and an improved data base for meteorological modelling.

Fuel dynamics

Fuel accretion over time in the absence of fire is an important characteristic of mediterranean ecosystems. An understanding of fuel dynamics over time is thus essential to development of a prescribed fire programme. As the general flammability increases over time, a recurrence interval for fire (or other manipulative technique) can be calculated to maintain low fuel loads, simulate a natural vegetation mosaic, or meet some other land management objective.

Economic studies

In the past, prescribed fire has been simultaneously characterized as being one of the least expensive and one of the most expensive techniques to get a particular job done. Both judgments have elements of truth, but both may be inapplicable to most future situations. As prescribed fire becomes a flexible, widely used technique, more precise achievement of objectives will be obtained and more difficult situations may be approached using prescribed fire. Actual and relative costs of burning to achieve a given objective should be analyzed.

Refined smoke management systems

Prescribed burning is bound to become a more widely applied technique in wildland management. Fuel buildups in many mediterranean ecosystems will be removed and some of this fuel will become smoke. Research is needed on the long range transport of pollutants from burning and pollutant impact on visibility and climate. The interaction of wildland smoke with urban pollution is another relatively unknown but important research topic. Such research can lead to improved systems for wildland smoke dispersal consistent with social goals. In discussing this topic at the Stanford workshop, most participants believed the smoke issue to be much more important in the United States as compared to other mediterranean-climate areas.

NATURAL FIRE RESEARCH

Natural fire management is a valuable fire strategy where primeval ecosystem states are desirable. This strategy has recently been redefined as '<u>prescribed</u> natural fire' in order to reinforce the preplanned nature of the strategy and avoid 'let-burn' connotations. In this Technical Note, 'natural fire' will be used in place of the term 'prescribed natural fire' to avoid confusion with prescribed fire management, although the emphasis on planning for natural fires is an essential concept of the strategy. The research priorities in natural fire management can be grouped into three general categories:
- frequency, intensity, and extent of naturally-occurring fires;
- changes in natural systems since fire exclusion became effective and means of reversing these changes;
- definition of impacts that may become externalized and of mitigating measures for such impacts (smoke, for example).

Natural fire frequency

Biologists have long recognized the major historical influence of fire on mediterranean ecosystems, but there is little specific data on the frequency, intensity, and extent of fire before the introduction of fire suppression. In those ecosystems that might be considered for natural fire management or which may contain information suitable for transfer to such areas, historical fire data are a high priority. Recent studies in the United States by Arno and Sneck (1977) in Montana, McNeil (1975) in Oregon, and Byrne, Michaelsen and Soutar (1977) in California cover techniques that can be applied in mediterranean areas where natural process management is contemplated.

While the old saying 'lightning never strikes the same place twice' is still popular, it is somewhat inapplicable to characterizing natural fire frequency. Historical records in many areas indicate definitely higher lightning ignition frequency in certain areas than in others. Searching historical records and analyzing more recent ones may help to characterize natural ignition patterns. Fossil charcoal studies may shed light on long-term ignition frequencies, including fires set early in human history.

Social ignition patterns are not well understood. While there has been considerable speculation on the influence of pre-technological or aboriginal people on fire frequency, there exists little substantive data on the rôle these people played in mediterranean ecosystem dynamics. There remains substantial philosophical disagreement about what tools or technology differentiate 'early' people from 'modern' people. Dasmann (1975) defines the two as 'ecosystem' people and 'biosphere' people; the former live or lived within one or several ecosystems, totally dependent on that system for survival, while the latter draw their support from many systems. A recent specific definition for a Montana wilderness classified 'modern' people as using repeating firearms, phosphorus matches, and crosscut saws, clearly separating their ability to modify natural conditions from 'early' people without such tools (Gabriel 1976). The transition may have been much more subtle and gradual in other cultures. More definitions on a site-specific basis are necessary, especially where 'ecosystem' or 'early' people are considered to have been part of the system and where it is desirable to simulate their rôle in present day ecosystems.

Natural fire intensity and extent

These can be at least crude estimations of natural fire intensity and extent from fire scar data and vegetation patterns. The vegetation of many mediterranean ecosystems is destroyed by fire, is short-lived, or is affected by stem decay so that growth ring counts recording successive fires or delineating specific age-class patterns are difficult to obtain. The existing record in many areas, if significant, is slowly disappearing over time. Now is the time to salvage this information, for as fire use programmes evolve, historical fire records recorded by vegetation are bound to be reduced or eliminated. Where coniferous vegetation is present, this record may be several hundred years old, in North America easily predating 'modern', 'technological', or 'biosphere' people.

Ecosystem restoration

Another aspect of natural fire management is that most mediterranean ecosystems have been protected, not always successfully, from fire for many years. Some plant species have temporarily disappeared from the natural community. Stands may be decadent, filled with dead fuels. Where multi-storied canopies exist, tremendous changes in understory vegetation have occurred in some systems. What might have once been mosaics are not in some instances uniformly flammable vegetation types. Obviously, delicate management must restore a semblance of natural conditions before natural fires are allowed to run their course. This requires some idea of how the ecosystems have been altered over time and what management practices will be necessary to restore a mimic of the natural pattern. Some of the specific studies needed are covered in the other chapters of this Technical Note. Restorative techniques might include selective mechanical manipulation or prescribed fire. Allowing natural fires only during a restricted relatively moist season has also been used as a restorative technique (Gunzel 1974).

External impacts

The last group of natural fire research priorities are those relating to external impacts of such programmes. Even the most wild areas cannot be considered as ecological islands, and while some study has in the past been made of impacts on wild areas from surrounding areas, natural fire programmes can potentially have adverse impacts on surrounding areas. These impacts can often be minimized or essentially eliminated through proper planning of natural fire zones, while others may not be so easily controlled. Erosional impacts on downstream values after widespread intense fires may be a serious problem in some areas, especially where downstream urban communities are involved. Other impacts include smoke and the threat of fire passing from a natural fire zone into surrounding lands. In the California parks that now have natural fire zones, neither smoke nor fire escape has posed a problem, but in different vegetation and geographical areas these impacts might become significant. Smoke impacts to be addressed include potential quantity-quality aspects and how the absence of timing control interacts with other sources of natural or man-made air pollution. Fire escape considerations include necessary buffers for zone edges when vegetation is relatively continuous or extremely flammable.

Natural fire research has a rather limited audience, composed of those involved with perpetuating natural ecosystems. The areas most likely to receive such management in mediterranean-climate ecosystems are wilderness areas and national parks. Sequoia and Kings Canyon National Parks in California, and Kosciusko National Park in Australia are areas where natural fire research as outlined above could be conducted and compared.

BIBLIOGRAPHY

AGEE, J.K. 1974. *Environmental impacts from fire management alternatives*. National Park Service, Western Regional Office. San Francisco.

ARNO, S.F.; SNECK, K.M. 1977. *A method for determining fire history in coniferous forests of the mountain west*. USDA For. Serv. Gen. Tech. Rep. INT-42. Intermountain Forest and Range Experiment Station, Ogden, Utah.

BYRNE, R.; MICHAELSEN, J.; SOUTAR, A. 1977. Fossil charcoal as a measure of wildfire frequency in southern California: a preliminary analysis. In: *Proceedings of the Symposium on the Environmental Consequences of Fire and Fuel Management in Mediterranean Ecosystems*. H.A. Mooney and C.E. Conrad (Eds), p. 361- 367. USDA For. Serv. Gen. Tech. Rep. WO-3. US Department of Agriculture, Washington, D.C.

COUNTRYMAN, C. 1974. *Can southern California wildland conflagrations be stopped?* USDA For. Serv. Gen. Tech. Rep. PSW-7. Pacific Southwest Forest and Range Experiment Station, Berkeley, California.

DASMANN, R. 1975. *National parks, nature conservation, and 'future primitive'*. Paper prepared for South Pacific Conference on National Parks, February 1975. Wallington, New Zealand.

GABRIEL, H.W. 1976. *Wilderness ecology: the Danaher Creek drainage, Bob Marshall Wilderness, Montana*. Ph.D. dissertation, Univ. Montana, Missoula, Montana.

GUNZEL, L. 1974. National policy change - natural prescribed fire. *Fire Manage.*, 35, p. 6-8.

KESSELL, S. 1976. Wildland inventories and fire modelling by gradient analysis in Glacier National Park. *Proc. Tall Timbers Fire Ecol. Conf.*, 14, p. 115-162.

MCNEIL, R. 1975. *Vegetation and fire history of a ponderosa pine - white fir forest in Crater Lake National Park*. Mast. Sci. thesis, Oregon State Univ., Corvallis, Oregon.

PHILPOT, C. 1974. New fire control strategy developed for chaparral. *Fire Manage.*, 35, p. 3-7.

STOREY, T. 1972. FOCUS: A computer simulation model for fire control planning. *Fire Tech.*, 8 (2), p. 91-103.

Management problems and solutions at the interface between man and mediterranean wildlands

Carl C. Wilson

The hazardous mediterranean climate, highly flammable vegetation, and rugged terrain, all important elements of fire behaviour, become problems only in the presence of people. People recreate and build homes in the mediterranean wildlands because of the delightful climate and will continue to do so as long as space is available. People start most fires, and their mere presence tends to warp fire suppression strategies because fire agencies must protect lives and property threatened by fires rather than 'back off' and build firelines around fire perimeters.

International, interagency, and interdisciplinary teams can work together in the urban/wildland interface to reduce the potential for unwanted fires in the mediterranean environment. Most of the forest fires in both Mediterranean countries and in California are caused by people. About 20 to 30 per cent of the fires in California are caused by lightning. In contrast, it is rare for lightning to cause more than 2 to 3 per cent of the fires in the Mediterranean region. Many of the fires are categorized as 'unknown' or 'miscellaneous' in each region (Naveh 1973; Velez 1976).

In spite of recent technological advances, fire problems are still increasing. In the Mediterranean region, both the number of fires and area burned are increasing (FAO-UNEP 1975). In 1970, in Spain and the south of France, fires blackened an area of 160,000 ha; Italy lost 81,000 ha of forests, maquis, and garrigue in 1971. Le Houerou (1973) estimated that 10 per cent of the forests and shrublands in southeastern France, Corsica, Sardinia, Sicily, and northeastern Algeria are burned annually by wildfires, destroying 200,000 ha at an estimated direct cost of US $50 million. In the 1970 fire disasters in California (between September 22 and October 4), there were 733 separate fires burning nearly 235,000 ha of grass-, brush- and timber-covered wildlands. The fires completely destroyed 722 homes, killed 16 people, and caused an estimated US $233 million in damage.

We appear to be sliding back two steps for each step we take forward in solving the management problems in the interface. Yet, the situation is not hopeless. In the first place, forest fire agencies can give top priority to the prevention of man-caused fires. Secondly, more pilot projects like the Perimetre Pilote du Massif des Maures in the Department of Var in France and the Laguna-Morena R&D project in San Diego County, California could be established. Thirdly, forestry and fire agencies can work closely with the local residents in developing an environment which is fire-safe for them and their property. Finally, more joint symposia, such as the Stanford workshop, could be held to allow scientists from different countries to exchange views and co-ordinate research.

At the Stanford workshop, four problem areas were identified as high priority topics for future research: pilot projects, fire prevention, prescribed burning and innovative fuel management.

PILOT PROJECTS

Research results are usually produced and applied on a piecemeal basis. There are few examples which demonstrate the application of all existing and relevant fire management knowledge. This could be at the sites of pilot projects, such as Des Maures in the south of France and Laguna-Morena in San Diego County, California. Such efforts would help identify gaps in knowledge; studies of ecological effects of wildfires and prescribed burns could also be conducted in these areas.

One of the most effective pilot projects in integrated forest fire management is located in the region of Provence-Côte d'Azur in the south of France (Fig. 1). Susmel (1973) and Coquet (1975) report that the Massif Des Maures

Unesco, 1978. *Fire and fuel management problems in mediterranean-climate ecosystems: research priorities and programmes.* (MAB Technical Notes 11.)

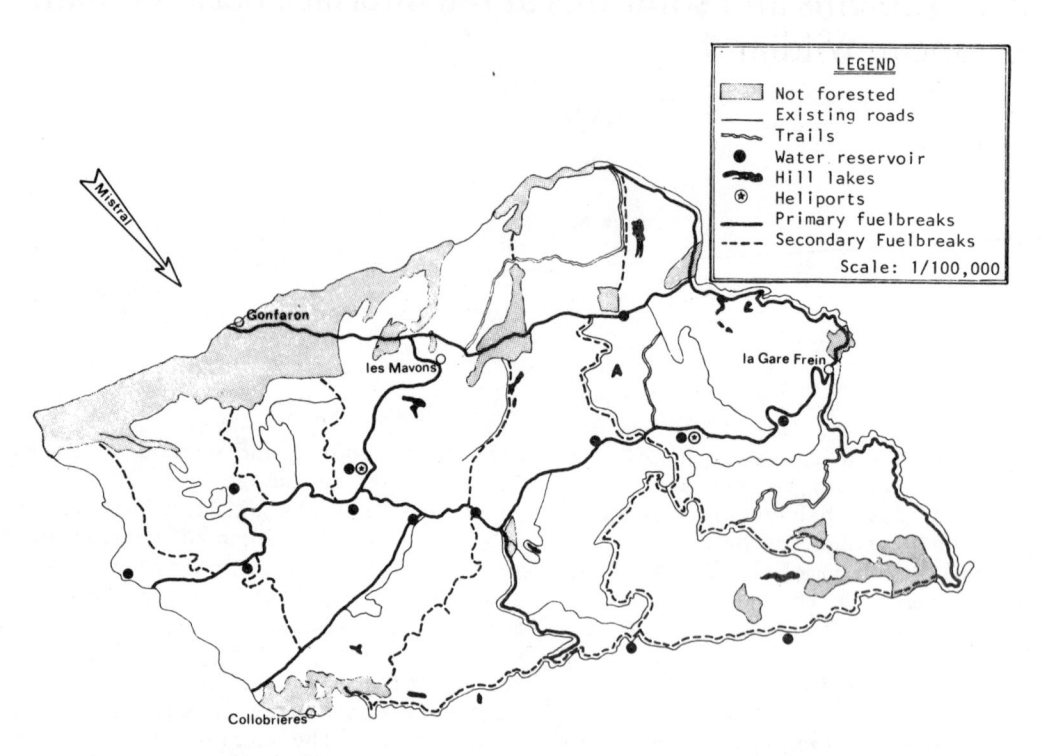

Figure 1. *Pilot district in fire management (Des Maures, France).*

project was selected because of a history of serious fires, and its topographic, ecological, and land ownership characteristics. The area is 19,000 ha in size of which 16,000 ha are covered with trees and shrubs. Elevation varies from 70 m to 780 m above sea level. The country is moderately steep, and the forest vegetation consists of sweet chestnut, white oak, holm oak, cork oak, and stone pine. The mediterranean maquis, which Susmel (1973) calls the 'main tinderbox', is interpersed with the forest trees.

The plan of work was prepared with the help of many specialists, scientists and technicians, with strong participation of the local residents. The primary objectives were: to apply new principles and methods of fire control, and to conduct experiments in integrated land management and test new techniques and organizations. Since 1965, the area has been broken into pre-attack-planned compartments; roads and heliports have been built and water tanks installed. The surface of some heliports serve as catchment areas for the nearby water tanks. Motorized prevention patrols cover 96 to 144 km each day, and aerial detection is activated when the mistral (foehn winds) is predicted. Special information and education efforts have been directed toward the local population.

By 1976, 94 km and 940 ha of primary fuelbreaks had been installed and maintained. This is about 98 per cent of the planned total. These are 100 m wide, dividing the area into six sections of 3,000 to 5,000 ha each. The secondary fuel-breaks, which are about 40 km in length and 50 to 100 m wide, break the unit into compartments of 1,000 ha each. A third kind of fuel-break consists of replacing the maquis and smaller pines with low flammability sweet chestnuts, red oaks, walnuts, tree of heaven, and fig trees. A fourth kind of fuel modification consists of fruit-growing, either non-irrigated, such as vineyards, or irrigated orchards. The public land for these purposes is located on the windward (north) side of the district and is leased to the farmers. Finally, some of the fuel-breaks are grazed by sheep and/or goats under careful management.

A pilot project similar to the one in France has been established in southern California. The Laguna-Morena Demonstration Area covers 15,000 ha in San Diego County (Fig. 2). A large variety of vegetation exists in the area including desert scrub, chamise, manzanita, grasses, ceanothus, live oak, and timber. The age of the vegetation varies from two years to over thirty since the last fire. The 1966 soil survey covers nearly all of the area.

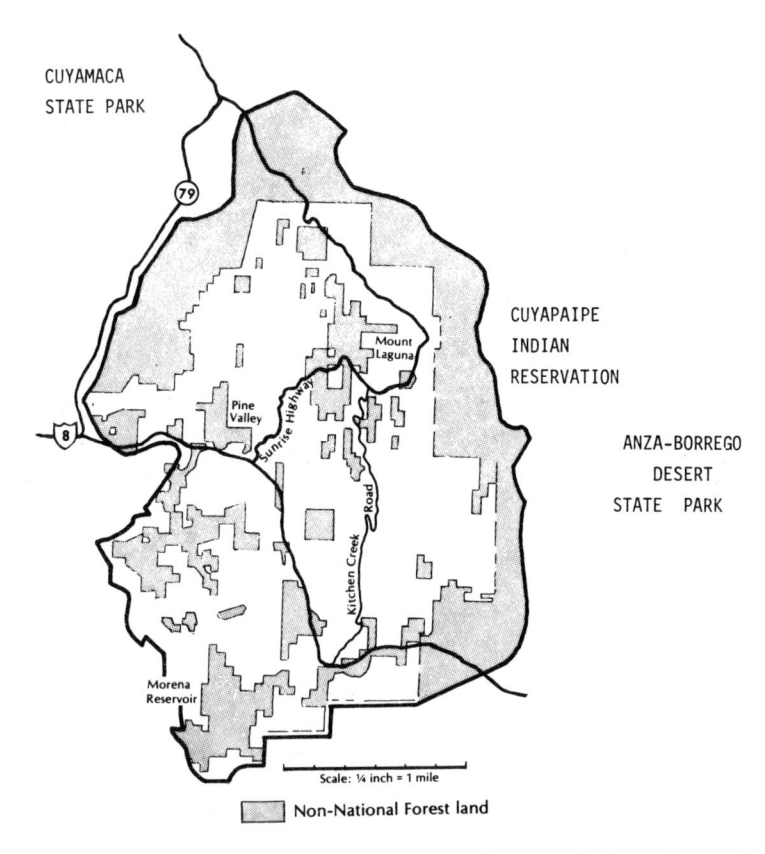

Figure 2. *Vegetation management pilot project and demonstration area (Laguna-Morena, California, USA).*

A large portion of area is under public ownership and is administered by the Forest Service, Bureau of Land Management, Cuyamaca State Park, and Bureau of Indian Affairs. This arrangement not only allows for freedom to experiment and work with few restrictions but also encourages intergovernmental co-operation. In addition, there is a great deal of information on fuel management and type conversion available for this part of California. Also, there are high value recreation areas within and adjacent to the pilot district.

The proposed objectives of the demonstration area are to: reduce potential for catastropic wildfires; involve all government agencies, private groups, and private individuals in integrating fire into wildland management; demonstrate and monitor proved chaparral management techniques; develop guidelines for chaparral management which can be used elsewhere; develop new ideas and new techniques of vegetation management.

In the meantime, a Research and Development Programme was chartered in early 1976 by the Forest Service for southern California. The mission of the programme is: to develop, test, and demonstrate vegetation management plans, techniques, and systems designed to maintain or enhance productivity of chaparral and related lands. The programme will be composed of a core research work unit assigned to a programme manager and his staff. A key aspect of the programme will be on-the-ground demonstrations and evaluations. The programme manager will work directly with land managers to accomplish this and to keep programme activities relevant to real-world problems.

Co-ordinated wildland fuels management programmes for mediterranean climates should recognize all tools available, including fire, to achieve defined objectives. These programmes must address the consequences of both proposed and existing management policies and the liability problems encountered by those who attempt management in a system of changing land use patterns and changing values. Finally, such a programme must recognize the public responsibility for sharing in the management costs and risks of wildland fuels management on all lands, public and private, when the public will benefit.

Participants at the Stanford workshop recommended that governments in mediterranean climates designate pilot areas throughout the world to demonstrate and implement management programmes for review of the rôle of fire in ecosystems and resource management.

FIRE PREVENTION

People cause most of the destructive fires in countries with mediterranean climates. It costs far less to prevent a forest fire than to fight one. Techniques to aid in preventing most man-caused fires are readily available in some countries but not in others. Hence, there are two urgent needs. First, fire prevention needs to be given high priority in each country. Second, there needs to be a sharing among nations of information on available fire prevention techniques, including educational materials, mass media and hazard reduction guides. Unless unwanted fires are prevented, the freedom to conduct desirable prescribed burning projects will be sharply limited. Many of the tools needed to prevent fires are already available. The major need is to disseminate knowledge of these fire prevention aids to all countries faced with these problems.

PRESCRIBED BURNING

Numerous seminars, conferences, and workshops have been conducted on the desirability, practice, and benefits of prescribed burning in the United States and Australia. Yet, little attention has been given to the technique of prescribed burning in mediterranean climates where the risk of escapes is the greatest. Also, in the United States there have been national and regional training sessions in the behaviour of wildfires for about 20 years, but few have dealt specifically with the behaviour of prescribed fires. The Stanford workshop participants concurred in the need for an international training workshop on the fundamentals of fire behaviour of prescribed fires.

INNOVATIVE FUEL MANAGEMENT

Recent fire disasters in California in the man/wildlands interface have shown that conventional hazard reduction and fuel modification methods can fail under critical weather conditions. An innovative kind of fuel management system is needed which can widen the interface, to provide a buffer and prevent disasters such as those which occurred in Santa Barbara in July 1977, October 1971, September 1964 and March 1961.

Domesticated plants, such as avocados, citrus, grapes, and olives might be established and cultivated under permit in a zone (possibly 32 km long and .30 km wide) above the residences and other valuable structures. In the long run, the costs to the local people and the general public would be drastically reduced. A similar approach is being used on the north side of the Des Maures pilot district in France to minimize the impact of the mistral (strong, dry north winds). The primary objective would be to prevent disasters such as those noted above in Santa Barbara.

The green belt concept is not new in the United States. Early in 1970, the US Forest Service, in co-operation with the California Division of Forestry, Lake Arrowhead Sanitation District, and the University of California at Riverside began to develop an irrigated green belt in Maloney Canyon (San Bernardino National Forest, California). Later, after the destructive fires of 1970, the Task Force on California's Wildland Fire Problem recommended the development of green belts 'for the purpose of reducing the damage to lives and property from wildland fires.'

Neither is the green belt idea new in the Mediterranean Basin. French specialists have attempted to integrate fire knowledge in the 19,000 ha Des Maures 'pilot district'. Not only have they broken the area into 1,000 ha compartments by primary and secondary fuel-breaks, but they have also dedicated about 7,500 ha of public lands on the northern side of the Des Maures District to vineyards and orchards to serve as a buffer zone under critical weather conditions.

Conceptually, managers of wildland areas know what physical techniques are available to reduce probability of fire escapes from wildland areas. What seems to be lacking is a clear understanding of what land managers can do to limit damage outside the wildlands in case of escape. For example, all but 75 ha of the Santa Barbara fire burned through a neighbourhood located in fire hazardous, subdivided brushland. Some possible approaches might be: (1) altering zoning regulations; (2) special building codes; (3) involving local residents in the land use planning process; (4) developing a clear-cut understanding by residents of the hazardous environment on the 'man side' of the interface; (5) altering the expectations of the residents that their homes can be saved under critical weather conditions; (6) establishing and clarifying liability for loss of lives and property in the interface; and (7) special taxation for fire protection in the interface.

In certain countries, it would be useful for forest and/or wildland managers to try to influence the behaviour of people adjacent to the wildlands through action at the local government or administrative level. Participants at the Stanford workshop suggested that all too often, foresters and other resource managers tend to shy away from political and social problems, and that consequently, zoning and land use planning 'battles' are lost by default. Various techniques of local involvement have been used in some countries with varying degrees of success depending on local laws, culture, etc. However, very little information about these techniques is available. An inventory and case study analyses would help resource managers select the approaches or methods most applicable to their specific management problems in the man/wildland interface.

BIBLIOGRAPHY

COQUET, J.C. 1975. Périmètre pilote du massif des Maures. In: *Numéro Spinal de la revue forestière française*, p. 376-384. Les Incendies de Forêts. Tome 2.

FAO-UNEP. 1975. *Detection and control of forest fires for the protection of the human environment. Proposals for a global programme.* Carl C. Wilson, Consultant. Rome.

LE HOUEROU, H.N. 1973. Fire and vegetation in the Mediterranean Basin. *Proc. Tall Timbers Fire Ecol. Conf.*, 11, p. 237-277.

NAVEH, Z. 1973. The ecology of fire in Israel. *Proc. Tall Timbers Fire Ecol. Conf.*, 11, p. 131-170.

SUSMEL, L. 1973. *Development and present problems of forest fire control in the Mediterranean region.* FAO, Rome.

VELEZ, R.M. 1976. Personal communication, 29 September.

Facilitating communication between researcher, manager and the public

Alan R. Taylor and Robert Z. Callaham

Several levels of information transfer failure in fire management hamper the planning, co-ordination, and support of fire programmes in mediterranean-climate ecosystems. The first level is the transfer of information to the public. Governments and other institutions in mediterranean ecosystems do not effectively communicate to the public the rôle of fire in those ecosystems.

The second level of information transfer failure occurs between fire managers and researchers, causing several important problems:
- Users often do not communicate their most important problems to researchers.
- Researchers sometimes conduct projects not related to management problems.
- Research results may be presented in a form not readily interpretable by land managers and owners.
- Potential users frequently do not know how, or are reluctant, to incorporate new information into active management programmes.
- Because of differences in mission, priorities, and objectives of various fire and fuel management agencies and land owners, research results often are applied differently, producing disagreement among those who use these results.

The third level of information transfer failure is that which occurs among researchers. It has several causes:
- regional differences in philosophy, procedures and methodology;
- lack of shared information on methodology, equipment and models;
- incompatibility of data and terminologies;
- serious delays before new results and models appear in the literature;
- frequently unfulfilled need for continuous training and recycling of scientists;
- lack of ready access to the published literature;
- language barriers;
- gaps between basic and applied research.

Co-ordination in solving these problems can be the task of various international institutions, including Unesco, in particular through its intergovernmental Man and the Biosphere (MAB) Programme, the International Council of Scientific Unions' Scientific Committee on Problems of the Environment (SCOPE), the International Union of Forest Research Organizations (IUFRO), and the Regional Forestry Commissions of the Food and Agricultural Organization (FAO). However, national initiatives are also essential to solve this continuing problem, which is not unique to research and management in mediterranean-climate ecosystems.

INTERNATIONAL ORGANIZATIONS

The following sections present brief introductions to the organizations featured under this topic. For more information on the Unesco Programme on Man and the Biosphere (MAB) the reader should refer to Unesco publications such as the MAB Technical Notes, MAB Report Series, the quarterly journal Nature and Resources; for information on the International Union of Forestry Research Organization (IUFRO), see Speer (1972); for a recent discussion of the North American Forestry Commission's Fire Management Study Group (FAO), see 'Minutes, Tenth Meeting, Fire Management Study Group, North American Forestry Commission.'[1]

Unesco-MAB

The general objective of the programme is to develop, within the natural and social sciences, a basis for the rational use and conservation of the resources of the biosphere and for the improvement of the global relationship between

---

1. Unpublished report on file at Northern Forest Fire Laboratory, USDA, Forest Service, Missoula, Montana 59806, USA.

man and his environment; to predict the future consequences of today's actions, and thereby to strengthen man's ability to manage the natural resources of the biosphere (Unesco 1972). To meet this objective, the International Coordinating Council for MAB is implementing 14 international interdisciplinary scientific projects through MAB National Committees in Unesco's member states. MAB Project 2, dealing with the ecological effects of land use and management practices on temperate and mediterranean forest landscapes, includes research on the ecological impact of forest fires (Unesco 1975).

SCOPE

The Scientific Committee on Problems of the Environment is a component of the International Council of Scientific Unions (ICSU). Established in 1969, SCOPE objectives are to: advance knowledge of the influence of man on his environment and the effects of environmental changes upon man; and serve as an interdisciplinary council of scientists to advise on environmental problems of major international importance. SCOPE Project 2, 'Dynamic changes and evolution of ecosystems', has begun an analysis of the ecological effects of fire (Munn and Cain 1977). The symposium, 'Environmental consequences of fire and fuel management in mediterranean climate ecosystems', of which SCOPE was a co-sponsor, is an example of SCOPE Project 2 activities.

IUFRO

The Union, founded in 1892, promotes collaboration among scientists in more than 350 forestry research organizations in some 86 countries of the world (Speer 1972). IUFRO is organized in six major divisions. Major co-operation among forest fire scientists is centered within Division 1, Site and Silviculture, in Subject Group $1.09, Forest Fire Research. The subjects of information systems and terminology are treated in Division 6, Subject Group $6.03.

FAO Regional Commissions

The purposes of the Regional Forestry Commissions of FAO are to: (1) advise the Assistant Director-General of the Forestry Department of FAO on formulation of forest policy, (2) review and co-ordinate policy implementation at the regional level, (3) exchange information, and (4) advise on suitable practices and actions on technical problems. There are six Regional Forestry Commissions under FAO: Asia-Pacific, Latin American, North American, Near East, European, and African. At present, only the North American Forestry Commission includes a fire management study group. The primary objective of this study group is to exchange ideas and information and promote mutual assistance among the participating countries of the North American Forestry Commission. The study group is made up of a committee on fire control, one on fire prevention, and one on fire research. The three committees review the activities of member nations and make recommendations to the Fire Management Study Group for submission to the North American Forestry Commission.

PROPOSED ACTIVITIES

Participants at the Stanford workshop proposed the following activities to enhance communication.
1. To facilitate *public* acceptance and understanding of the rôle of fire in mediterranean-climate ecosystems:
   - environmental education programmes (through national initiatives and through the Unesco-UNEP Environmental Education Programme);
   - mass media utilization (through national initiatives);
   - interpretive programme (through national initiatives).
2. To improve communications, co-operation and understanding between *researchers* and *fire managers*:
   - improved summarization, review and interpretation of the state of fire and fuel management practices (national initiatives);
   - preparation of guidelines, handbooks, and training materials on desirable management practices and technology (national initiatives);
   - convening of workshops and seminars on desirable fire and fuel management practices (FAO and national initiatives);
   - development of exchange programmes for fire management personnel-managers assigned to research, researchers to management (FAO and national initiatives);
   - utilization of existing and development of new co-ordination mechanisms such as councils, boards and committees (national initiatives);
   - development of demonstration programmes and areas that utilize advanced fire and fuels management techniques (FAO and national initiatives);
   - intensified development of advanced systems for using and managing fire and fuels, including resource inventory and

information systems, modelling techniques, and fire behaviour and effects simulation capabilities (national initiatives);
- creation of small information analysis centres composed of managers and researchers responsible for technology transfer in the areas of the above recommendations (national initiatives).
3. To improve information transfer among *researchers*:
  - multilingual vocabularies, terminologies, thesauri, and glossaries (IUFRO $1.09, IUFRO $6.03, national initiatives);
  - translation services (national initiatives);
  - bibliographic data bases (FAO, IUFRO, national initiatives);
  - access to numerical data sets, including response coefficients, inventories, and other resource data (national initiatives);
- standardization of methodologies (IUFRO, SCOPE, Unesco-MAB);
- communications among scientists through seminars, workshops, and newsletters (IUFRO, SCOPE, Unesco-MAB);
- international exchange of scientists (Unesco-MAB, national initiatives);
- utilization of existing current research information systems (FAO, Unesco-MAB, national initiatives).

To facilitate communication between researcher, manager and the public efforts must be made at both the national and international levels. International progress cannot be made without national progress, and national efforts can more fully succeed with international co-operation. The similarities of information transfer problems between mediterranean-climate countries make co-operative efforts efficient and valuable means of achieving fire management objectives.

BIBLIOGRAPHY

MUNN, R.E.; CAIN, C. 1977. SCOPE: The environmental voice of world science. *Environmental Sci. and Tech.*, 11 (12), p. 1056-1060.

SPEER, J. 1972. *IUFRO. 1892-1972.* International Union of Forestry Research Organizations, Norway.

UNESCO 1972. *Final report of the International Coordinating Council of the Programme on Man and the Biosphere (MAB). First Session.* MAB Report Series No. 1. Unesco, Paris.

UNESCO 1975. *Report of an expert panel on MAB Project 2: Ecological effects of different land use and management practices in temperate and mediterranean forest landscapes.* MAB Report Series No. 19. Unesco, Paris.

Unesco publications: national distributors (Abridged list)

Argentina	EDILYR, Tucumán 1699 (P.B. 'A'), 1050 Buenos Aires.
Australia	*Publications*: Educational Supplies Pty. Ltd., Post Office Box 33, Brookvale 2100, N.S.W. *Periodicals*: Dominie Pty. Subscriptions Dept., P.O. Box 33, Brookvale 2100, N.S.W. *Sub-agent*: United Nations Association of Australia (Victorian Division), 2nd Floor, Campbell House, 100 Flinders Street, Melbourne 3000.
Austria	Dr. Franz Hain, Verlags- und Kommissionsbuchhandlung, Industriehof Stadlau, Dr. Otto-Neurath-Gasse 5, 1220 Wien.
Brazil	Fundaçao Getúlio Vargas, Serviço de Publicaçoes, caixa postal 9.052-ZC-02, Praia de Botafogo 188, Rio de Janeiro (GB); Carlos Rohden, Livros e Revistas Técnicos Ltda., Av. Brigadeiro Faria Lima 1709, 6º. andar, caixa postal 5004, Sao Paulo.
Burma	Trade Corporation no. (9), 550-552 Merchant Street, Rangoon.
Canada	Renouf Publishing Company Ltd., 2182 St. Catherine Street West, Montreal, Que. H3H 1M7.
Chile	Bibliocentro Ltda., Constitución nº 7, Casilla 13731, Santiago (21).
Cuba	Ediciones Cubanos, O'Reilly Nº. 407, La Habana.
Czechoslovakia	SNTL, Spalena 51, Praha I (Permanent display); Zahranicni literatura, 11 Soukenicka, Praha I. *For Slovakia only*: Alfa Verlag, Publishers, Hurbanova nam. 6, 893 31 Bratislava.
Denmark	Ejnar Munksgaard Ltd., 6 Nørregade, 1165 København K.
Ecuador	RAYD de Publicaciones, Garcia 420 y 6 de Diciembre, Casilla 3853, Quito. Casa de la Cultura Ecuatoriana, Núcleo del Guayas, Pedro Moncayo y 9 de Octubre, casilla de correos 3542, Guayaquil.
Ethiopia	Ethiopian National Agency for Unesco, P.O. Box 2996, Addis Ababa.
Finland	Akateeminen Kirjakauppa, Keskuskatu 1, SF-00100 Helsinki 10.
France	Librairie de l'Unesco, 7, place de Fontenoy, 75700 Paris. CCP Paris 12598-48.
German Democratic Republic	Buchhaus Leipzig, Postfach 140, 701 Leipzig or international bookshops in the German Democratic Republic.
Germany (Fed. Rep.)	S. Karger GmbH, Karger Buchhandlung, Angerhofstr. 9, Postfach 2, D-8034 Germering/Munchen. *For scientific maps only*: GEO Center, Postfach 800830, 7000 Stutgart 80. *For 'The Courier' (German edition only)*: Colmanstrasse 22, 5300 Bonn.
Ghana	Presbyterian Bookshop Depot Ltd., P.O. Box 195, Accra; Ghana Book Suppliers Ltd., P.O. Box 7869, Accra; The University Bookshop of Ghana, Accra; The University Bookshop, of Cape Coast; The University Bookshop of Legon, P.O. Box 1, Legon.
Greece	International bookshops (Eleftheroudakis, Kauffman, etc.).
Hong Kong	Swindon Book Co., 13-15 Lock Road, Kowloon. Federal Publications (HK) Ltd, 5A Evergreen Industrial Mansion, 12 Yip Fat Street, Wong Chuk Hang Road, Aberdeen.
Hungary	Akadémiai Könyvesbolt, Váci u. 22, Budapest V; A.K.V. Könyvtárosok Boltja, Népköztársaság utja 16, Budapest VI.
Iceland	Snæbjörn Jónsson & Co. H.F., Hafnarstraeti 9, Reykjavik.
India	Orient Longman Ltd, Kamani Marg, Ballard Estate, Bombay 400 038; 17 Chittaranjan Avenue, Calcutta 13; 36a Anna Salat, Mount Road, Madras 2; B-3/7 Asaf Ali Road, New Delhi 1; 80/1 Mahatma Gandhi Road, Bangalore-560001; 3-5-820 Hyderguda, Hyderabad-500001. *Sub-depots*: Oxford Book & Stationery Co., 17 Park Street, Calcutta 700016; Scindia House, New Delhi 110001; Publications Section, Ministry of Education and Social Welfare, 511 C-Wing, Shastri Bhaven, New Delhi 110001.
Indonesia	Bhratara Publishers and Booksellers, 29 Jl. Oto Iskandardinata III, Jakarta; Gramedia Bookshop, Jl. Gadjah Mada 109, Jakarta. Indira P.T., Jl. Dr. Sam Ratulangi 37, Jakarta Pusat.
Iran	Iranian National Commission for Unesco, Avenue Iranchahr Chomali No. 300, B.P. 1533, Tehran; Krarazmie Publishing and Distribution Co., 28 Vessal Shirazi Street, Shahreza Avenue, P.O. Box 314/1486, Tehran.
Ireland	The Educational Company of Ireland Ltd, Ballymount Road, Walkinstown, Dublin 12.
Israel	Emanuel Brown (formerly Blumstein's Bookstores), 35 Allenby Road *and* 48 Nachlat Benjamin Street, Tel Aviv; 9 Shlomzion Hamalka Street, Jerusalem.
Jamaica	Sangster's Book Stores Ltd., P.O. Box 366, 101 Water Lane, Kingston.
Japan	Eastern Book Service Inc., C.P.O. Box 1728, Tokyo 10092.
Kenya	East African Publishing House. P.O. Box 30571, Nairobi.
Korea (Republic of)	Korean National Commission for Unesco, P.O. Box Central 64, Seoul.
Libyan Arab Jamahiriya	Agency for Development of Publication and Distribution, P.O. Box 34-35, Tripoli.
Madagascar	Commission nationale de la République démocratique de Madagascar pour l'Unesco, Boîte postale, 331, Tananarive.
Malaysia	Federal Publications, Lot 8323 Jalan 222, Petaling Jaya, Selangor.
Malta	Sapienzas, 26 Republic Street, Valletta.
Mexico	SABSA, Insurgentes Sur nº. 1032-401, Mexico 12 D.F.
Netherlands	N.V. Martinus Nijhoff, Lange Voorhout 9, 's-Gravenhage; Systemen Keesing, Ruysdaelstraat 71-75, Amsterdam 1007.
Netherlands Antilles	Van Dorp-Eddine N.V., P.O. Box 200, Willemstad, Curaçao, N.A.
New Zealand	Government Printing Office, Government bookshops: Mulgrave Street, Private Bag, Wellington; Rutland Street, P.O. Box 5344, Auckland; 130 Oxford Terrace, P.O. Box 1721, Christchurch; Alma Street, P.O. Box 857, Hamilton; Princes Street, P.O. Box 1104, Dunedin.
Nigeria	The University Bookshop of Ife; the University Bookshop of Ibadan, P.O. Box 286; The University Bookshop of Nsukka; The University Bookshop of Lagos; the Ahmadu Bello University Bookshop of Zaria.
Norway	*Publications*: Johan Grundt Tanum, Karl Johans gate 41-43, Oslo 1. *For 'The Courier'*: A/S Narvesens Litteraturtjeneste, Box 6125, Oslo 6.
Pakistan	Mirza Book Agency, 65 Shahrah Quaid-e-azam, P.O. Box 729, Lahore 3.
Philippines	The Modern Book Co., 926 Rizal Avenue, P.O. Box 632, Manila D-404.
Southern Rhodesia	Textbook Sales (PVT) Ltd., 67 Union Avenue, Salisbury.
Singapore	Federal Publications (S) Pte Ltd., No. 1 New Industrial Road, off Upper Paya Lebar Road, Singapore 19.
South Africa	Van Schaik's Bookstore (Pty.) Ltd., Libri Building, Church Street, P.O. Box 724, Pretoria.
Spain	MUNDI-PRENSA LIBROS S.A., apartado 1223, Castelló 37, Madrid 1; Ediciones LIBER, Apartado 17, Magdalena 8, Ondárroa (Vizcaya); DONAIRE, Ronda de Outeiro, 20, apartado de correos 341, La Coruña; Librería AL-ANDALUS, Roldana, 1 y 3, Sevilla 4; Librería CASTELLS, Ronda Universidad 13, Barcelona 7.
Sudan	Al Bashir Bookshop, P.O. Box 1118, Khartoum.
Sweden	*Publications*: A/B C.E. Fritzes Kungl. Hovbokhandel, Regeringsgatan 12, Box 16356, S-103 27 Stockholm. *For 'The Courier'*: Svenska FN-Förbundet, Skolgränd 2, Box 150 50, S-104 65 Stockholm (Postgiro 18 46 92).
Switzerland	Europa Verlag, Rämistrasse 5. 8024 Zürich; Librairie Payot, 6, rue Grenus, 1211 Genève 11.
Thailand	Suksapan Panit, Mansion 9, Rajdamnern Avenue, Bangkok; Nibondh & Co. Ltd., 40-42 Charoen Krung Road, Siyaeg Phaya Sri, P.O. Box 402, Bangkok; Suksit Siam Company, 1715 Rama IV Road, Bangkok.
Uganda	Uganda Bookshop, P.O. Box 145, Kampala.
U.S.S.R.	Mezhdunarodnaja Kniga, Moskva G-200.
United Kingdom	H.M. Stationery Office, P.O. Box 569, London SE1 9NH; Government bookshops: London, Belfast, Birmingham, Bristol, Cardiff, Edinburgh, Manchester.
United Republic of Tanzania	Dar es Salaam Bookshop, P.O. Box 9030, Dar es Salaam.
United States	Unipub, 345 Park Avenue South, New York, NY 10010.

A complete list of distributors is available from the Office of Publications, Unesco

[B.] SC.78/XXIX.11/A

Based on an international symposium held at Stanford in August 1977, this Technical Note discusses the need for a more complete understanding of the dynamics of fire-affected ecosystems of mediterranean regions so that management policies can be developed and assessed on a rational basis. Recommendations are made for facilitating communication between researchers and resource managers facing similar problems in different parts of the world and mechanisms suggested for translating research findings into practical management programmes. Specific aspects of fire and fuel management covered include the role of species characteristics; mineral and sediment pools and fluxes; community productivity, structure and diversity; assessment of alternative fire management policies; and the interface between man and mediterranean wildlands.

A stylized 'ankh', the ancient Egyptian sign for life, has been incorporated into the symbol of the Programme on Man and the Biosphere (MAB).

ISBN 92-3-101688-1